T0144164

# Intelligent Video Surveillance Systems

## An Algorithmic Approach

# Intelligent Video Surveillance Systems

## An Algorithmic Approach

Maheshkumar H. Kolekar

## CRC Press
Taylor & Francis Group
Boca Raton  London  New York

CRC Press is an imprint of the
Taylor & Francis Group, an **informa** business
A CHAPMAN & HALL BOOK

CRC Press
Taylor & Francis Group
6000 Broken Sound Parkway NW, Suite 300
Boca Raton, FL 33487-2742

© 2018 by Taylor & Francis Group, LLC
CRC Press is an imprint of Taylor & Francis Group, an Informa business

No claim to original U.S. Government works

Printed on acid-free paper
Version Date: 20180529

International Standard Book Number-13: 978-1-4987-6711-8 (Hardback)

This book contains information obtained from authentic and highly regarded sources. Reasonable efforts have been made to publish reliable data and information, but the author and publisher cannot assume responsibility for the validity of all materials or the consequences of their use. The authors and publishers have attempted to trace the copyright holders of all material reproduced in this publication and apologize to copyright holders if permission to publish in this form has not been obtained. If any copyright material has not been acknowledged please write and let us know so we may rectify in any future reprint.

Except as permitted under U.S. Copyright Law, no part of this book may be reprinted, reproduced, transmitted, or utilized in any form by any electronic, mechanical, or other means, now known or hereafter invented, including photocopying, microfilming, and recording, or in any information storage or retrieval system, without written permission from the publishers.

For permission to photocopy or use material electronically from this work, please access www.copyright.com (http://www.copyright.com/) or contact the Copyright Clearance Center, Inc. (CCC), 222 Rosewood Drive, Danvers, MA 01923, 978-750-8400. CCC is a not-for-profit organization that provides licenses and registration for a variety of users. For organizations that have been granted a photocopy license by the CCC, a separate system of payment has been arranged.

**Trademark Notice:** Product or corporate names may be trademarks or registered trademarks, and are used only for identification and explanation without intent to infringe.

---

**Library of Congress Cataloging-in-Publication Data**

---

Names: Kolekar, Maheshkumar H., 1972- author.
Title: Intelligent video surveillance systems : an algorithmic approach / Maheshkumar H Kolekar.
Description: First edition. | Boca Raton, Florida : CRC Press/Taylor & Francis Group, [2019] | Includes bibliographical references and index.
Identifiers: LCCN 2018008209| ISBN 9781498767118 (hardback : acid-free paper) | ISBN 9781315153865 (ebook)
Subjects: LCSH: Video surveillance. | Image analysis--Data processing. | Artificial intelligence.
Classification: LCC TK6680.3 .K65 2019 | DDC 621.389/28--dc23
LC record available at https://lccn.loc.gov/2018008209

---

**Visit the Taylor & Francis Web site at**
**http://www.taylorandfrancis.com**

**and the CRC Press Web site at**
**http://www.crcpress.com**

Dedicated to my father Hanmant Baburao Kolekar, who believed in me and motivated me to pursue a career in teaching.

# Contents

# III  Surveillance Systems                                             139

# List of Figures

# List of Tables

# Foreword

"We will be remembered only if we give to our younger generations,
a prosperous and safe India"

Dr. A. P. J. Abdul Kalam
Honorable Former President of India

In recent years, the scourge of terrorism has been affecting the whole world.
India has suffered a number of casualties as a result of bomb blasts in market
areas, indiscriminate firing in crowded places and suicide attacks. The terrorist attack on the Parliament in December 2001 and later in Mumbai on
November 26, 2008 have resulted in intensifying counter-terrorism operations
to thwart such acts in the future. In order to effectively counter the terrorist
operations, there is the need for accurate and timely intelligence commonly
known as actionable intelligence, or, as I would like call it, predictive intelligence.

There are different types of intelligence inputs, which are needed to form
a complete and reliable information package on which the security agencies
can take pre-emptive action. These can be communication intercepts, photo
imagery, human-sourced inputs, etc. However, in proximity to or within important area such as official premises and airports, video surveillance has become
a very important tool to counter terrorists or anti-social acts. While many
cities in the country have established video-based networks as part of Smart
City projects, the suggestions from this book will greatly help the city ad-
ministration in many ways, such as control of traffic, crowd control and crime
control and detection, to name a few.

With the recently developed IP-based systems and smart cameras, which
also have filters for human facial and behavioral identification, the intelligent
video surveillance will go a long way in enhancing various applications. Un-
doubtedly, these require complex algorithmic solutions and are a daunting
task for scientists. Dr. Maheshkumar Kolekar has therefore taken a vital step
in writing this book, explaining certain concepts and solutions as well as giving illustrations, figures, and photographs to explain his hypothesis.

I am therefore delighted to introduce the book, *Intelligent Video Surveillance Systems*, written by Dr. Maheshkumar Kolekar. I am very happy that as
a follow-up to the research project being undertaken by him at IIT Patna, he
has decided to compile the details into a book. It is also noteworthy that this

book on video surveillance is coming out in the early stages of the field, which is under rapid development. This book therefore can play a significant role in bringing about the understanding and subsequent maturity of the field.

Though many research papers are available, they provide limited perspectives about the field, its application potential, and the techniques. This book gives a broad view about the available techniques, implementation aspects, emerging techniques, and algorithms. I have particularly liked the idea that instead of searching for concepts from a variety of papers, the author has attempted to introduce and compile different implementation details, which will allow ease in further research.

Over the past few years, the author was working on a project, "Abnormal Human Activity Recognition to Enhance Security," which was sponsored by the principal scientific advisor to the government of India. Being a member of the project monitoring committee, I had the opportunity to evaluate his project work. Since this topic is very important from a security point view, I have always encouraged Dr. Kolekar to continue his research work in this field and asked him to motivate more students to work in this area. As a result of that he has initiated an elective course on intelligent video surveillance in IIT Patna and I am happy to know that this is now a very popular course in the institute. I came to know that many students of different disciplines such as electrical engineering, computer science and engineering, and mechatronics engineering joined this course in 2018. Seeing this trend, I am sure that in the coming years, this course and resultant research will become very popular in other technical institutes and universities, and this book will be very useful for such courses. Also, I urge the faculty members of other institutes to initiate similar courses, since this is a very important area in today's scenario.

These days, many public or open places are equipped with cameras to monitor the security of that area. This increased adaptation of security cameras to survey public areas has caused a digital data explosion that human operators have difficulty keeping up with. Even though humans are the most intelligent creatures in this world, there are many shortcomings in human-operated systems. At the moment, the best solution to attain a safe environment requires automatic tools for human operators to monitor the digital camera images. Due to the fatigue factor, human observers cannot pay full attention to several video screens for periods of long duration. In order to support human operators, automatic abnormal human activity recognition and tracking systems are introduced in the surveillance video for security in public places such as railway stations, airports, and banks.

Normally, for working on an intelligent video surveillance systems course, understanding of digital image processing, digital video processing, and machine learning are pre-requisites. But to allow easier adaptation, the author has devoted the first and second chapters to digital image processing and digital video processing, respectively, and explains machine-learning algorithms such as HMM with implementation details in a chapter on human activity recognition to make this book self-sufficient. The flow of the chapters is also logical. The first two chapters are on basic concepts in image and video processing, to create background for the readers in this area. Chapter 3 explains background modelling, which is the first building block of any surveillance system. Object classification, human activity recognition, and video object tracking are discussed in Chapters 4, 5, and 6, respectively. Chapter 7 describes various camera models used for surveillance, and in the final Chapter 8, the author discusses some techniques and applications that have been implemented in various domains.

The chapters in the book describe various machine learning concepts and demonstrate practical applications of advanced techniques, such as hidden Markov models, support vector machine, convolutional neural network, and deep learning. Instead of explaining many concepts in brief, the author has tried to explain a few important concepts at length, with required implementation details for surveillance application. I am sure that by reading this book, readers can easily implement various algorithms and understand video surveillance-related concepts. I wish Dr. Maheshkumar Kolekar, the author of this book, fellow researchers working in related areas, and other readers of the book further success in their endeavors in enhancing better security and to be of service to humanity.
Jai Hind.

Air Marshal Bhushan Gokhale (Retired)
PVSM, AVSM, VM
26 January, 2018

# *Preface*

Nowadays, visual surveillance is going to be an important tool for making the world more secure. Everybody wants to secure their places of work and residence, and entrepreneurs are also ready to invest in making such surveillance systems. The objective of this book is to present the recent progress that has been made in video surveillance systems, and enlist the latest techniques and algorithms that have been developed. Human activity recognition is an important building block of any video surveillance system. It will help us to recognize humans and describe their actions and activities, from information acquired by video cameras. This is a really challenging task and researchers are proposing many new techniques and algorithms for it.

Although there are few books dealing with various aspects of video surveillance, a book comprehensively covering the image processing, video processing, algorithm, and implementation aspects of video surveillance is urgently needed. This book covers video surveillance systems including background modeling, foreground detection, blob segmentation, activity recognition, individual behavior recognition, group behavior recognition, and tracking of human objects. Chapters 1 and 2 introduce basic concepts of image processing and video processing, respectively.

Chapter 3 focuses on background modeling, which is the first building block of any conventional video surveillance system. This chapter also includes shadow removal and foreground extraction. Chapter 4 describes techniques of object classification, which include recent techniques based on deep convolutional neural network and region-based convolutional neural network models. Chapter 5 tackles human activity recognition problems in detail, exploring various techniques including dynamic time warping. This chapter also covers a few classifiers based on machine learning techniques, such as hidden Markov model and support vector machine. From all recognized human activities, abnormal activities will be classified using some domain knowledge. Chapter 6 considers more complex scenarios of tracking multiple objects and handling challenges such as occlusion of objects. This includes classification of multiple interacting objects from videos. Multi-object tracking scenarios like handling of partial views and occlusions will be done using camera networks, as described in Chapter 7. Persons observed by a multi-camera system may appear in different views simultaneously or at different times depending on the overlap between views. Chapter 7 describes different types of surveillance cameras. Chapter 7 also addresses implementation aspects of surveillance systems, such as camera network, placement, camera calibration, camera communication, camera coordination, and cooperation. Chapter 8 covers the implementation aspect of video surveillance systems and emerging techniques in visual surveillance systems for various applications such as baggage exchange detection, and fence crossing detection. Chapter 8 also explains the use of surveillance systems for military and transportation applications.

This book can be utilized in several ways depending on the course level and desired learning objectives. Chapters 1 to 6 are sufficient for an undergraduate course on video surveillance. Chapters 3 to 8 are sufficient for a postgraduate course for the students with image and video processing backgrounds. One can design a postgraduate course for those students who do not have image and video processing backgrounds by including Chapters 1 to 8. Chapters 5 to 8 are very useful for PhD research scholars, scientists, and engineers who wish to understand algorithms and implementation in surveillance. I hope that the design methodologies, algorithms, and applications discussed throughout this book will provide guidelines for researchers and developers working on vision-based applications. This book contains 18 algorithms and their implementation details with results. This book can also serve as a reference book for undergraduate and graduate students with interests in other courses such as computer vision, machine learning, image processing, and video processing.

I would like to mention that I became interested in this area because of Professor Somnath Sengupta (IIT Kharagpur), Professor K. Palaniappan (University of Missouri, Columbia, MO), and Air Marshal B. N. Gokhale P.V.S.M., A.V.S.M., V.M. (Retd). I would like to mention thanks to the Honorable R. Chidambaram, Principal Scientific Advisor to the Government of India, for sanctioning my project related to video surveillance, which created my interest in this area and motivated me to write this book. I would like to thank Shri G. P. Srivastava, Scientific Consultants PSA office, and Shri Neeraj Sinha, Scientist G, PSA office, for their support and encouragement. I would like to mention my sincere thanks to Professor Pushpak Bhattacharya, Director IIT Patna, for his valuable suggestion in implementing machine learning algorithms such as hidden Markov model-based classifiers.

I would like to thank Professor P. K. Biswas (IIT Kharagpur), Professor A. K. Roy (IIT Kharagpur), Professor Subhasis Chaudhuri (IIT Bombay, Mumbai), Professor Vikram Gadre (IIT Bombay, Mumbai), Professor A. N. Rajgopalan (IIT Madras, Chennai), and Shri J. K. Mukherjee (BARC Mumbai) for their support in the valuable discussion about many concepts explained in this book. I acknowledge the solid support received for data recording and testing from the students of IIT Patna, especially Pranjali Manesh Kokare, Prateek Sharma, Deepanway Ghosal, Shobhit Bhatnagar, Durgesh Kumar, Garima Gautam, Himanshu Rai, Kanika Choudhary, and Deba Prasad Dash. I extend my special thanks to Aastha Sharma, senior acquisitions editor, CRC Press, Taylor & Francis Group, New Delhi, who helped me at every stage of the development process of the book.

I would like to thank my mother Sushila H. Kolekar, my wife Priti, and daughters Samruddhi and Anwita, for their constant support and encouragement. Today the dream of my mother came true.

Dr. Maheshkumar H. Kolekar
Associate Professor, EE Dept.
IIT Patna

# Part I

# Basics of Image and Video Processing

# Chapter 1

## Basics of Image Processing

Visual information has a bigger impact on people. Nowadays, with the advent of digital technologies, producing digital images and videos is becoming increasingly easy. Since the digital image and video databases are of huge size, there is a need to develop tools to assist in the synthesis and analysis of digital image and video sequences.

The aim of this chapter is to provide an overview of the existing techniques in digital image processing to the reader. In this chapter, an introduction to digital image processing (DIP), the DIP system, important components of DIP, commonly used methods for DIP, image segmentation, and some applications of DIP are discussed. Digital video is just a sequence of images, and hence most of the image processing techniques can be extended to video processing. Additional video processing techniques mainly involve motion analysis and video compression, which will be discussed in the next chapter.

## 1.1 Introduction to Digital Image Processing

With the advent of digital technologies, acquiring digital images is becoming increasingly easy. Images can be acquired by using a variety of devices such as still cameras, video cameras, x-ray devices, ultrasound, electron microscopes, and radar. Image processing is used for a variety of purposes, including entertainment, education, medical, business, industrial, military, surveillance, and security. The goal in each case is to extract useful information from the image.

### 1.1.1 Why Digital Image Processing?

Images are avaliable everywhere because of the fast development in image acquisition tools. Nowadays, many mobile handsets come with image and video recording facility. Hence, there is a need to process these images and videos to get more information. Digital image processing is motivated by the following applications:

*Improvement of pictorial information for human perception*: This type of application involves improving the quality of images for human perception. Typical applications are noise filtering, contrast enhancement, deblurring, and remote sensing.

*Image processing for autonomous machine applications*: This has various applications in industry, such as quality control and assembly automation.

*Efficient storage and transmission*: To process the image so that the required disk space for storing the image can be reduced. It is possible to compress the image so that it can be transmitted with low bandwidth channel.

### 1.1.2    What Is Digital Image?

The term *image* is a 2-D light intensity function $f(x, y)$, where $x$ and $y$ denote spatial coordinates and $f$ denotes an intensity value at that point. If we discretize an image $f(x, y)$ both in spatial coordinates and intensity, a digital image will be formed. A *digital image* is simply a matrix whose row and column indices identify the coordinates of a point in the image, and the corresponding matrix element value identifies the gray level at that point. The element of such a digital image is called a *pixel*, which is an abbrevation of a picture element. Intensity values of pixels are defined by bits. Since 8-bit intensity range has 256 possible values, 0 to 255 is called gray scale. RGB color images use 8-bit intensity ranges for each color. Since RGB images contain $3 \times 8$ bit intensities, they are also referred to as 24-bit color images.

Image size should not be confused with the size of the real-world representation of an image. Image size specifically describes the number of pixels within a digital image. The real-world representation of a digital image requires one additional factor called resolution. Resolution is the spatial scale of the image pixels. For example, an image of $3300 \times 2550$ pixels with a resolution of 300 pixels per inch (ppi) would be a real-world image size of $11'' \times 8.5''$.

### 1.1.3    What Is Digital Image Processing?

Digital image processing uses computer algorithms to perform some operations on an image to extract some useful information from it. In DIP, input is an image and output is a processed image. As shown in Figure 1.1, DIP includes acquisition of input image and processing an image to deliver a modified image. Here, an image of pears is segmented based on the color of each pear (see ebook for color).

Digital image processing is preferred because it has many advantages over analog image processing. DIP avoids problems such as noise and signal distortion. It allows the use of matrix operations to be applied to the input image. Hence, DIP offers better performance and better implementation methods

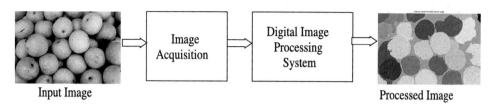

Input Image                                    Processed Image

**FIGURE 1.1**: Digital image processing system

that would not be possible with analog processing.

## 1.2 Digital Image Processing System

The elements of a general purpose system capable of performing the image processing operation are acquisition, storage, processing, communication, and display.

### 1.2.1 Image Acquisition

This is the first step of a DIP system, because without an image, no processing is possible. To capture an image, we use a digital camera. After illumination of the object of interest, the light reflected from the object has to be captured by the camera. The light reflected from the object of interest is focused by some image sensor so that the camera can record the scene. An image sensor consists of a 2-D array of cell, and each cell represents a pixel. An image sensor is capable of converting incident light into a voltage, which in turn is converted into a digital number. More incident light means higher voltage, and higher intensity value at that point. When the camera is capturing an image, incident light is allowed to enter for a small duration called an exposure time. During this exposure time, charges start accumulating in each cell, which is controlled by the shutter. If the exposure time is too small or too large, the result is an underexposed or overexposed image, respectively. Digital images can also be obtained by conversion of analog images into digital images with a scanner. There are some advanced methods of image acquisition such as three-dimensional (3-D) image acquisition. The 3-D image acquisition uses two or more cameras that have been aligned precisely around a target to create a 3-D or stereoscopic scene. Some satellites use 3-D image acquisition techniques to build accurate models of different surfaces.

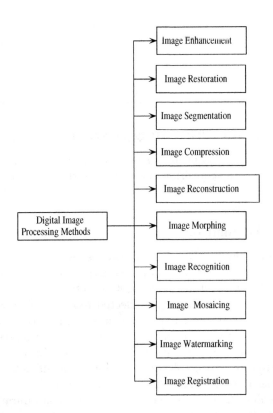

**FIGURE 1.2**: Digital image processing methods

## 1.2.2 Storage

The storage of image information is very important for its long-term preservation. The recommended storage medium is the hard drive. Hard drives are viable for 3 to 10 years, with an average of about 5 years. Images can be compressed to reduce their size for storage and increase the transmission speed. Image compression techniques are broadly classified into lossless and lossy. Lossless compression file formats save some space without changing the image data. Lossy compression saves some space at the cost of data. JPEG is a well-known format of lossy image compression. Lossy compression should be used whenever the storage space is one of the more important constraints than loss of some original data. Lossy compression is most effective at reducing file size and increasing transmission speed.

## 1.2.3 Processing

An image is processed by using some mathematical operations or signal processing techniques, or a computer algorithm, to generate the modified output image or parameters related to an image. Since the image is a two-dimensional matrix of intensity values, standard signal-processing or statistical techniques are applied. The computer system processes the image using some algorithms and generates the output image. Adobe Photoshop is also an example of image processing.

## 1.2.4 Communication

Since image size is usually large, the key consideration in the transmission of an image is bandwidth. If we want to transmit an image over a low bandwidth channel, we first compress the image to reduce its size and then decompress the image at the receiver end to reproduce the original input. An image usually contains lot of redundancy that can be exploited to achieve compression. Lossless compression is used for medical imaging and legel documents. Lossless compression reduces the size of the image without errors, but only to a certain extent. Beyond this point, errors are introduced, and such compression is called lossy compression. JPEG is one of the popular lossy image compression formats. Since in image compression a small loss in quality is usually not noticeable, lossy methods are used for storing and transmitting natural images such as photographs.

## 1.2.5 Display

For displaying digital images, flat TV-screens, or computer monitors, or printing devices are used. Early printers, such as dot matrix and daisywheel

(a)                                   (b)

**FIGURE 1.3**: (a) Original image, (b) Segmented image

printers, were called impact printers because they operated by striking an ink ribbon against the paper. The printer that uses non-impact technology, such as ink-jet or laser technology, provides better quality at higher speed than impact printers.

## 1.3   Digital Image Processing Methods

Figure 1.2 shows the commonly used methods for processing the digital image. In this section, we will discuss those methods in brief.

### 1.3.1   Image Enhancement

Enhancement means improving the quality of an image for some specific application by some *ad-hoc* technique. It improves the appearance of an image by increasing the dominance of some features. It should be noted that enhancement of certain features can be acheived at the cost of suppressing others. An example of image enhancement is noise removal using media filtering.

### 1.3.2   Image Restoration

Restoration is an attempt to estimate the original image by applying effective inversion of the degrading phenomenon. This method requires knowledge of the degradation phenomenon. An example of image restoration is removing motion blur in an image of a fast-moving object.

### 1.3.3 Image Segmentation

The segmentation process usually refers to making different parts of an image. The segmentation renders an image into more meaningful parts, which helps for analysis of the image. It is mainly used to extract objects or regions in images. In image segmentation, a label is assigned to every pixel based on some visual characteristics, such as color, intensity, texture, or depth. The pixels with the same label constitute a region. Hence, the result of image segmentation is a set of regions. The different regions can be different objects in the image and background. Hence, all the different regions collectively cover the entire image. If segmentation is done well, then all sccessive stages in image analysis are made simpler.

Since image segmentation has wide applications, several algorithms have been developed. For image segmentation, usually two approaches are followed, namely discontinuity based and similarity based. In the discontinity-based approach, an image is partitioned into parts based on discontinuties in intensity values. These discontinuties are at the boundaries of objects. Hence, edge detection operators are used to detect edges, and these edge pixels are linked properly to identify the object boundaries. Canny edge detection and Laplacian edge detection techniques are used in this approach.

In the similarity-based approach, we assume that the pixels of the same object have some similarity. The image is partitioned into different regions that are similar according to some predefined criteria. The examples of this approach are thresholding, region growing, and region splitting and merging. Practically, the intensity values of pixels of the same object change slightly because of variation in illumination and noise. Hence, to improve the segmentation performance, these techniques are modified based on domain knowledge of a particular application.

### 1.3.4 Image Compression

Image compression is used to reduce the amount of memory required to store a digital image and transmit it at low bandwidth. It is used in image transmission applications which include broadcast television, remote sensing via satellite, and other long-distance communication systems. Image storage is required for several purposes, such as document, medical images, magnetic resonance imaging (MRI), radiology, and motion pictures. Image-compression algorithms are divided into lossy or lossless. In lossless compression, information content is not modified. Compression is achieved by reducing redundancy and this is used for archival purposes, as well as for medical imaging and legal

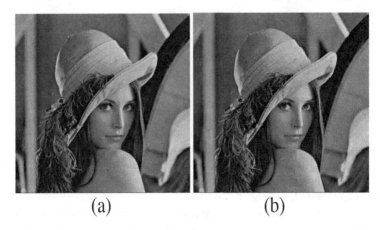

(a)                                    (b)

**FIGURE 1.4**: (a) Original image, (b) Compressed JPEG image (27:1)

documents. In lossy compression, information content is reduced and it is not recoverable. Lossy compression methods are epecially used at low bit rates, which introduce compression artifacts. Lossy methods are especially used for natural images such as photographs. Figure 1.4 shows the original image and JPEG compressioned image with compression ratio 27:1. Though image 1.4 (b) is compressed 27 times the size of the original image, its quality is sufficiently good.

### 1.3.5    Image Reconstruction

Image reconstruction is a technique used to create 2-D and 3-D images from sets of 1-D projections. It is used in the common imaging modalities, such as CT (computed tomography), MRI (magnetic resonance imaging), and PET (positron emission tomography). These modalities are very useful in the medicine and biology field. The CT image is generated by using several images that are acquired from x-ray projection data at different angles around the patient. In image reconstruction, image quality is based on radiation dose. Images of acceptable quality can be reconstructed at lower radiation doses. Image reconstruction is performed using two types of methods, namely analytical reconstruction and iterative reconstruction.

#### 1.3.5.1    Analytical Reconstruction

The most commonly used analytical reconstruction methods on commercial CT scanners are all in the form of filtered back-projection (FBP). It uses a 1-D filter on the projection data before back-projecting (2-D or 3-D) the data onto the image space. This is widely used on clinical CT scanners be-

**FIGURE 1.5**: Three frames are shown from a sequence morphing from child to woman. The middle image represents the morphing process.

cause of its computational efficiency and numerical stability. Flohr et *al.* have presented a review of analytical CT image reconstruction methods [14].

#### 1.3.5.2 Iterative Reconstruction

Nowadays, iterative reconstruction is prefered over conventional FBP techniques because of the following advantages:

1) In iterative reconstruction, important physical factors such as focal spot and detector geometry, photon statistics, x-ray beam spectrum, and scattering can be incorporated more accurately.

2) Interative reconstruction produces images with lower noise and higher spatial resolution.

3) As compared to FBP, beam hardening artifacts, windmill artifacts, and metal artifacts are reduced in iterative reconstruction.

4) As per recent clinical studies, iterative reconstruction algorithms have demonstrated radiation dose reduction upto 65% compared with the FBP-based reconstruction approach.

5) The apperance of images from iterative reconstruction may look different, since there is an intrinsic difference in data handling between FBP-based reconstruction and iterative reconstruction.

### 1.3.6 Image Morphing

Morphing is a process of transforming one image to another through a seamless transition. It is a technique that allows to blending two images, creating a sequence of *in-between* pictures. When played, this sequence converts the first image into the second, creating animation effect. This type of special effect is used to depict some technological processes and fantasy sequences. Figure 1.5 shows morphing of child to woman. Morphing algorithms are used to create convincing slow-motion effects by morphing between each individual frame using optical flow technology. Morphing techniques are used to enhance multimedia projects, presentations, movies, education, and training.

As the morphing proceeds, the first image is gradually faded out, while the second image is gradually faded in. The *in-between* images contain both images according to some percentage. The middle image contains fifty percent of the source as well as the destination image. It involves the image processing techniques of cross-fading and warping. The correspondence between the source and destination image is developed using points and lines. Sometimes features such as intensities and optical flow are used to find correspondence between the two images. For image morphing, algorithms such as mesh warping, field morphing, radial basis functions, thin plate splines, energy minimization, and multilevel free-form deformations are used.

### 1.3.7   Image Recognition

Image recognition determines whether the image contains some specific object or not. This task is very easy for humans, but is very difficult for machines. The existing algorithms are only for specific objects such as human faces, vehicles, animals, geometric objects, or characters. Some of the recognition problems are as follows:

*Object recognition*: In this problem, one or several prespecified objects can be recognized with their 2-D positions in the image or 3-D poses in the video.

*Identification*: An individual instance of an object is recognized, such as identification of the face of a specific person, or identification of a specific vehicle.

*Detection*: The specific object is detected in the image or the video sequence. Examples are detection of possible abnormal cells in medical images, and detection of a vehicle in a road-traffic monitoring system.

Several specialized tasks based on recognition are as follows:

*Content-based image retrieval*: This is a technique of finding all the images from the larger dataset of images that have a specific content. For example we can make a query such as, "show me all images similar to image X." Then the CBIR module displays all images similar to X by finding similarities relative to the target image.

*Pose estimation*: This is a technique of estimating the position or orientation of a specific object relative to the camera. For example, the application of assisting a robot arm in retrieving objects from a conveyor belt.

*Optical character recognition (OCR)*: This is a technique of identifying printed or handwritten text characters in images. It is used to encode the text in a format such as ASCII. This format can be easily edited.

### 1.3.8   Image Mosaicing

Image mosaicing is a method of creating one image by combining information from many images, for better understanding. The amount of data collected by the camera is very small compared to what is seen by the human

**FIGURE 1.6**: A photographic mosaic of fields and the growing condition of crops, acquired by drone

eye. Nowadays, there are special cameras with complex lenses to increase the field of view. However, we can increase the field of view of a normal camera by combining information from many images. Image mosaicing enables us to combine together many small images into one large image. The term photomosaic means compound photographs created by stitching together a series of adjacent pictures of a scene. Figure 1.6 shows a typical example of the mosaic of the pictures of fields and growing condition of crops acquired by drone for agricultural purpose. Many products are already available in the digital photography market to allow photos or even a video stream from a handheld camera to be stitched together into a wide-field mosaic. The technique is used for the applications in video compression, digital steadycam, digital video editing, and matte placement software.

### 1.3.9 Image Watermarking

Image watermarking is the technique of embedding information into a digital image in order to attach ownership. Watermarking is performed in such a way that it is difficult to remove. If the image is copied, then the watermark information is also carried with the copy. If the watermark information is visible in the image, it is called visible watermarking. Usually the information is text or a logo that indicates the ownership of a particular group. An example of a visible watermark is the logo at the corner of transmitted video included by a television broadcaster to inform viewers about the owner of the media. In invisible watermarking, information is added as digital data to an image, but it cannot be perceived as such. In steganography applications, the invisible watermark contains secret messages embedded in the digital signal.

### 1.3.10    Image Registration

Image registration is the technique of transforming different sets of data into one coordinate system. The different sets of data may be multiple photographs, data from different sensors, data acquired at different times, or data acquired from different viewpoints. It is necessary to integrate the data obtained from these different measurements. Image registration is used for various applications, such as medical image registration.

For monitoring tumors, images of the patient are acquired from different viewpoints. These images are registered in order to combine images from different viewpoints. This helps doctors with disease dignosis. In astrography, different images of space are registered for analysis purposes. The computer performs transformations on one image to make major features align with a second image using some control points. Image registration is used to create panoramic images. Many different techniques of image registration can be implemented in real-time devices such as cameras and camera phones.

## 1.4    Digital Image Segmentation

This is the technique of partitioning a given digital image into meaningful parts with respect to a particular application. Image segmentation is used to locate objects in images. Image segmentation is mainly used for analysing images. In other words, it is the process of assigning a label to every pixel in an image based on certain visual characteristics. The set of different regions generated by image segmentation collectively cover the entire image. Each pixel in the segmented region is similar to other pixels in the region with respect to some characteristic, such as color, intensity, texture, and depth.

For understanding any image, digital image segmentation is an important step. It is used to identify objects in a scene and extract their features, such as size and shape. In object-based video compression, image segmentation is used to identify moving objects in a scene. To enable mobile robots to find their path, image segmentation is used to identify objects that are at different distances from a sensor using depth information.

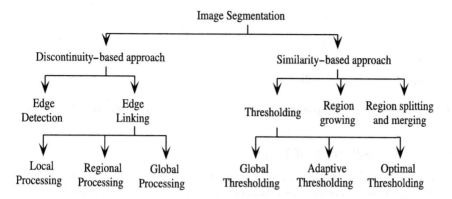

**FIGURE 1.7**: Classification of image segmentation approaches

## 1.4.1   Classification of Image Segmentation Techniques

The image segmentation approaches are broadly classified as *similarity-based approaches* and *discontinuity-based approaches* as shown in Figure 1.7. The similarity-based approach is based on partitioning an image into regions that are similar according to predefined criteria. This approach can be further classified as *thresholding, region growing,* and *region splitting and merging.* Region in an image is a group of connected pixels with similar properties. The thresholding-based approach can be further classified as global thresholding, adaptive thresholding, optimal thresolding, and local thresholding.

The discontinuity-based approach is to partition an image into different regions based on abrupt changes in intensity, such as edges in an image. In the edge detection-based approach, the edges are identified to detect the region or object boundaries. Detected edges are linked together to improve the performance. Edge linking techniques can be further classified as local processing, regional processing, and global processing. The Hough transform-based global processing method is widely used for edge linking.

## 1.4.2   Edge Detection

Edge detection is the process of detecting sharp discontinuities in an image. The discontinuities are abrupt changes in pixel intensity that characterize boundaries of objects in the image. Since edges often occur at object boundaries, this is used for object extraction. The edges can be extracted by convolving the image with the edge detection operator. Some edge detection operators are the Sobel, Canny, Prewitt, and Roberts operators.

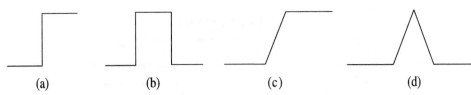

**FIGURE 1.8**: (a) Step edge, (b) Line edge, (c) Ramp edge, (d) Roof edge

### 1.4.2.1 Classification of Edges

Pixels at which the intensity of an image function changes abruptly are called edge pixels. These edge pixels can be detected by computing intensity differences in local image regions. Edges can be broadly classified into step edge, line edge, ramp edge, and roof edge, as shown in Figure 1.8. The step edge has a perfect transition from one segment to another. In the case of step edge, the image intensity abruptly changes from one side of the discontinuity to the other side. If a segment of an image is narrow, it necessarily has two edges in close proximity. This arrangement is called a line edge. Ramp edge occurs when there is smoother transition from one region to other. It is useful for modeling the blurred edge. Two nearby ramp edges in a line structure are called a roof.

### 1.4.2.2 Gradient Operator

The gradient of an image $f$ at location $(x, y)$ is given as:

$$\nabla f = grad(f) = \begin{bmatrix} g_x \\ g_y \end{bmatrix} = \begin{bmatrix} \frac{\partial f}{\partial x} \\ \frac{\partial f}{\partial y} \end{bmatrix} \tag{1.1}$$

The magnitude of the gradient vector is given by

$$mag(\nabla f) = \sqrt{g_x^2 + g_y^2} \tag{1.2}$$

The direction of the gradient vector measured with respect to the x-axis is given by

$$\alpha(x, y) = tan^{-1} \begin{bmatrix} g_x \\ g_y \end{bmatrix} \tag{1.3}$$

Obtaining the gradient requires computing the partial derivatives $\partial f/\partial x$ and $\partial f/\partial y$ at every pixel location. The digital approximations of the partial derivatives about the pixel location $(x, y)$ are given as

$$g_x = \frac{\partial f}{\partial x} = \frac{\partial f(x, y)}{\partial x} = f(x + 1, y) - f(x, y) \tag{1.4}$$

**FIGURE 1.9**: Roberts mask

and

$$g_y = \frac{\partial f}{\partial y} = \frac{\partial f(x,y)}{\partial y} = f(x, y+1) - f(x,y) \tag{1.5}$$

The above two equations can be implemented by using the mask given in Figure 1.9 . When diagonal edge direction is of interest, a Roberts cross-gradient operator is used. Consider the $3 \times 3$ image region shown in Figure 1.10. The Roberts operator compute gradient uses the following equations:

$$g_x = \frac{\partial f}{\partial x} = (z_9 - z_5) \tag{1.6}$$

$$g_y = \frac{\partial f}{\partial y} = (z_8 - z_6) \tag{1.7}$$

Roberts masks are not symmetric about the center point. The smallest symmetric mask is of size $3 \times 3$. These masks are useful for computing the direction of an edge. The simplest $3 \times 3$ mask is the Prewitts mask, which is given as follows:

$$g_x = \frac{\partial f}{\partial x} = (z_7 + z_8 + z_9) - (z_1 + z_2 + z_3) \tag{1.8}$$

$$g_y = \frac{\partial f}{\partial y} = (z_3 + z_6 + z_9) - (z_1 + z_4 + z_7) \tag{1.9}$$

To give more emphasis on the central pixel, the Sobel operator uses a weight of 2 in the center coefficient, which provides smoothing. Sobel masks are preferable because they have better noise supression.

$$g_x = \frac{\partial f}{\partial x} = (z_7 + 2z_8 + z_9) - (z_1 + 2z_2 + z_3) \tag{1.10}$$

$$g_y = \frac{\partial f}{\partial y} = (z_3 + 2z_6 + z_9) - (z_1 + 2z_4 + z_7) \tag{1.11}$$

It is to be noted that the sum of all coefficients of any edge detection mask is zero. Hence, a response of mask is zero in areas of constant intensity. Computing the gradient magnitude using Equation 1.2 requires squares and square roots calculations, which pose a computational burden, so an equation used to approximate gradient values is

$$M(x,y) \approx |g_x| + |g_y| \tag{1.12}$$

It still preserves relative changes in intensity levels.

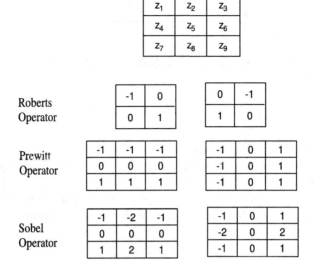

**FIGURE 1.10**: Image region of size 3 x 3 (z represents intensity values) and various edge operators

### 1.4.2.3    Laplacian Operator

The gradient operator discussed above detects the edges based on the first derivative of the image. The second derivative can be computed by differentiating the first derivative. The Laplacian operator searches for zero crossings in the second derivative of the image to find edges. Since the Laplacian of an image highlights regions of rapid intensity change, it is prefered than above other edge detectors. The shape of the Laplacian operator looks like Mexican hat as shown in Figure 1.11.

For images, there is a single measure, similar to the gradient magnitude, that measures the second derivative, which is obtained by taking the dot product of $\nabla$ with itself.

$$\nabla.\nabla = \begin{bmatrix} \frac{\partial}{\partial x} \\ \frac{\partial}{\partial y} \end{bmatrix} . \begin{bmatrix} \frac{\partial}{\partial x} \\ \frac{\partial}{\partial y} \end{bmatrix} \tag{1.13}$$

$$\nabla.\nabla = \frac{\partial^2}{\partial x^2} + \frac{\partial^2}{\partial y^2} \tag{1.14}$$

The operator $\nabla.\nabla = \nabla^2$ is called the Laplacian operator.
When the Laplacian operator is applied to the function $f$, we get

$$\nabla^2 f = \frac{\partial^2 f}{\partial x^2} + \frac{\partial^2 f}{\partial y^2} \tag{1.15}$$

The Laplacian operation can be expressed in terms of the difference equation given below

$$\frac{\partial f}{\partial x} = f(x+1, y) - f(x, y) \tag{1.16}$$

and

$$\frac{\partial^2 f}{\partial x^2} = f(x+1, y) - 2f(x, y) + f(x-1, y) \tag{1.17}$$

Also

$$\frac{\partial^2 f}{\partial y^2} = f(x, y+1) - 2f(x, y) + f(x, y-1) \tag{1.18}$$

This implies that

$$\nabla^2 f = [f(x+1, y) + f(x-1, y) + f(x, y+1) + f(x, y-1)] - 4f(x, y) \tag{1.19}$$

The masks corresponding to Equation 1.19 are shown in Figure 1.12 (a).

The Laplacian operator subtracts the intensity values of each of the neighboring pixels from the central pixel. When a discontinuity is present within the neighborhood of the pixel, the reponse of the Laplacian operator is a non-zero positive or negative value. The discontinuity will be in the form of point, line, or edge. One of the limitations of the Laplacian operator is that being an approximation to the second derivative, it is sensitive to noise in the image. To overcome this problem, the image is smoothed using a Gaussian filter before applying the Laplacian operators to reduce the high frequency noise components. Based on the pre-processing, the following Laplacian operators are developed:

1. Marr-Hildreth edge detector
2. Canny edge detector

### 1.4.2.4 Marr Hildreth Edge Detector

At an edge, sudden intensity change will give rise to a peak or trough in the first derivative and zero crossing in the second derivative. Marr and Hildreth argued that the intensity changes depend on image scale. Hence, we can use the operators of different sizes for edge detection. The large operators are used for blurry edges and small operators are used for sharp edges.

Marr and Hildreth proposed that the most satisfactory operator fulfilling the above-mentioned condition is the filter $\nabla^2 G$ with $G$ as a two-dimensional Gaussian function. The detailed steps of the algorithm are presented in Algorithm 1.

$$G(x, y) = e^{-\left(\frac{x^2+y^2}{2\sigma^2}\right)} \tag{1.20}$$

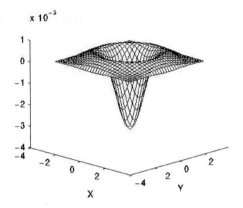

**FIGURE 1.11**: Laplacian operator (Mexican hat shape)

where $\sigma$ is the standard deviation.

$$\nabla^2 G = \frac{\partial^2 G}{\partial x^2} + \frac{\partial^2 G}{\partial y^2} \tag{1.21}$$

The Laplacian of Gaussian (LoG) is

$$\nabla^2 G = \frac{x^2 + y^2 - 2\sigma^2}{\sigma^4} e^{-(\frac{x^2+y^2}{2\sigma^2})} \tag{1.22}$$

The Gaussian part of $\nabla^2 G$ blurs the image, thus reducing the intensity of the structures, including noise. Gaussian is smooth in both the spatial and frequency domains, and thus is less likely to introduce ringing not present in the original image. $\nabla^2$ being isotropic, it avoids the use of multiple masks to calculate the strongest response at any point in the image.

Since both the Gaussian and Laplacian kernels are usually much smaller

---

**Algorithm 1** Marr Hildreth Edge Detection Algorithm

---

**Input:** Input image
**Output:** Image containing detected edges

1: Smooth the image using a Gaussian Low-pass filter to reduce error due to noise.
2: Apply 2-D laplacian mask shown in Figure 1.12 to the image. This Laplacian will be rotation invariant. It is often called the *Mexican Hat operator* because of its shape which is shown in Figure 1.11.
3: Loop through Laplacian and look for sign changes. If there is a sign change and the slope across this sign change is greater than some threshold, consider this pixel as an edge.

---

than the image, this method usually requires less computations. For real-time application, the LoG kernel can be calculated in advance so only one convolution needs to be performed at run-time on the image. If the threshold is zero, we may observe closed-loop edges. This is called the *spaghetti* effect.

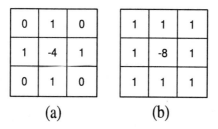

FIGURE 1.12: Laplacian masks

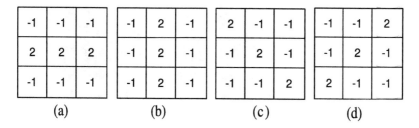

FIGURE 1.13: (a) Horizontal line mask, (b) Vertical line mask, (c) +45 degree line mask, (d) -45 degree mask

We can avoid this by using a positive threshold.

### 1.4.2.5 Isolated Point Detection

A Laplacian mask is used to detect an isolated point. The intensity at an isolated point will be quite different from its surroundings. Hence, the isolated point will be easily detectable by the Laplacian mask that is shown in Figure 1.12. If the absolute value of the response of the Laplacian mask at any point exceeds a specified threshold $T$, then that point is considered an isolated point. Such points are labeled 0, thus producing a binary image.

### 1.4.2.6 Line Detection

Figure 1.13 shows the line masks for horizontal line, vertical line, $+45^{o}$ line, and $-45^{o}$ line. Figure 1.14 shows the responses of these masks to the image shown in Figure 1.14(a).

### 1.4.2.7 Canny Edge Detector

The Canny edge detection algorithm is considered the optimal edge detector. Canny wanted to improve the performance of edge detectors available in the 1980s, and in 1986, he proposed an Algorithm 2 for edge detection that is very popular and efficient (see Figure 1.15). In it, the following list of criteria

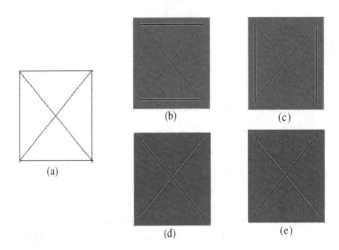

**FIGURE 1.14**: (a) Original image, (b) Response of horizontal line mask, (c) Response of vertical line mask, (d) Response of +45 degree line mask, (e) Response of -45 degree line mask

**FIGURE 1.15**: Photograph of John Canny

is used to improve the performance:

(1) Low Error Rate:

Probability of detecting real edge points should be increased, and probability of detecting non-edge points should be decreased.

(2) The edge points should be well localized:

The distance between the real edge pixels and the detected edge pixels should be at a minimum.

(3) Only one response to a single edge:

There should be only one response to one real edge pixel.

The Canny edge detector first smooths the image to reduce noise. Then it finds the image gradient to highlight regions with high spatial derivatives. After this, all those pixels in these regions that are not at maxima are supressed. Hysteresis is now used to further reduce the gradient array. Hysteresis is used to check the remaining pixels that have not been suppressed. Hysteresis uses two thresholds, and if the magnitude is below the low threshold ($T_1$), it makes the pixel as non-edge. If the magnitude is above the high threshold ($T_2$), it makes it edge. If the magnitude is between two thresholds, then it is set to zero unless there is a path from this pixel to a pixel which is an edge.

Other gradient-based algorithms like the Prewitt filter are not efficient

---

**Algorithm 2** Canny Edge Detection Algorithm

---

**Input:** Input image $f(x, y)$
**Output:** Image containing detected edges

1: Smooth the image using a Gaussian low-pass filter to reduce the noise.
2: Compute the gradient magnitude and angle for pixel where gradient has large magnitude to create edges.
3: Apply non-maxima suppression to the gradient magnitude image.
4: Use double thresholding and connectivity analysis to detect and link edges.
5: Edge tracking by hysteresis

---

when there is more noise. Moreover, parameters like kernel filter and coefficients are not adjustable. In contrast, the Canny edge detection algorithm is an adaptive algorithm in which parameters like standard deviation for the Gaussian filter and the threshold values are adjustable. In comparison with the Sobel, Prewitt, and Roberts operators, the Canny edge detection algorithm is computationally expensive. The Canny edge detection algorithm presented in Algorithm 2 performs better than all these operators, as shown in Figure 1.16.

## 1.4.3 Edge Linking

In the edge detection technique, sometimes the edge becomes discontinuous, and therefore we need a technique that can link these broken edges. In practice, noise and non-uniform illumination introduce spurious intensity discontinuities. Hence, the generated output has breaks in the boundary. As

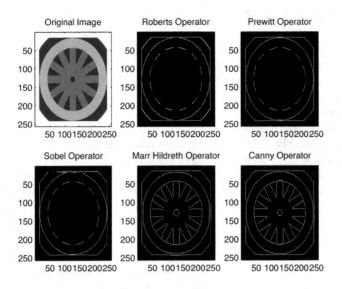

**FIGURE 1.16**: The figure indicates that the Canny detector performs better as compared to other detectors

a remedy, various techniques for linking edges are used.

### 1.4.3.1  Local Processing

For linking edge points, one of the easiest ways is to link pixels with similar characteristics. After the image had undergone edge detection, a small neighborhood ($3 \times 3$ or $5 \times 5$) of pixels is analyzed, and the similar pixels are linked. Similarity is established by two principle properties:

1. The strength of the response of the edge detection operator used to detect the edge pixel.

2. The direction of the operator.

For example, if we use gradient operator to produce the edge pixel, similarity can be deduced by:

For edge strength similarity

$$\left| \nabla f(x,y) - \nabla f(x^{'},y^{'}) \right| \leq T \tag{1.23}$$

where $T$ is the non-negative threshold.

For direction similarity

$$\left| \alpha(x,y) - \alpha(x^{'},y^{'}) \right| \leq A \tag{1.24}$$

where $\alpha(x, y)$ is the direction of the gradient, $\alpha(x, y) = tan^{-1} \frac{G_y}{G_x}$, and $A$ is an angle threshold.

### 1.4.3.2 Regional Processing

This technique is used for linking pixels on a regional basis, with the desired outcome being an approximation of the boundary of the region. Two approaches are used for regional processing, (1) functional approximation and (2) polygonal approximation. In functional approximation, we fit a 2-D curve to the known point. The main interest lies in the fast execution that yields an approximation to essential features of the boundary such as extreme points. Polygonal approximations can capture the essential shape features of a region, such as vertices of the polygon.

### 1.4.3.3 Global Processing Using Hough Transform

Local and regional processing are applicable in situations where knowledge about pixels belonging to individual objects is at least partially available. In regional processing, it makes sense to link a given set of pixels only if we know that they are part of the boundary of a meaningful region. If we have no knowledge about the object boundaries, all pixels are candidates for linking and thus have to be accepted or eliminated based on predefined global properties. In the global approach, we see whether sets of pixels lie on curves on a specified shape. These curves form the edges or region boundaries of interest. Global processing can be done by Hough transform. The Hough transform is a feature extraction technique used in image analysis. The basic idea of Hough transform is as follows:
1. Each straight line in Figure 1.17 can be described by an equation, $y = mx + c$.
2. Each white point, if considered in isolation, could lie on an infinite number of straight lines.
3. In the Hough transform, each point votes for every line it could be on.
4. The line with the most votes wins.

Any line can be represented by two numbers. We can also represent a line by using $(\rho, \theta)$.

$$\rho = x cos\theta + y sin\theta \tag{1.25}$$

Since we can use $(\rho, \theta)$ to represent any line in the image space as shown in Figure 1.18 (a), we can represent any line in the image space as a point in the plane defined by $(\rho, \theta)$, as shown in Figure 1.18 (b). This is called Hough space. The detailed algorithm of the Hough tranform is presented in Algorithm 3.

Some of the applications of the Hough transfrom in image analysis are as follows: The detection of groups of parallel lines can be used as important clues for text extraction. In applications involving autonomous navigation of air vehicles, Hough transform is used to extract two parallel edges of the principal runway. While analyzing soccer videos, the appearance of two or

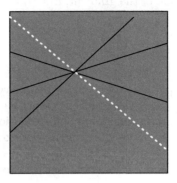

**FIGURE 1.17**: Each white point could lie on an infinite number of straight lines

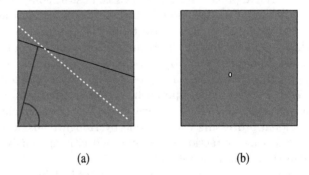

(a)                                          (b)

**FIGURE 1.18**: Each white point could lie on an infinite number of straight lines

---

**Algorithm 3** Hough Transform Algorithm

---

**Input:** Input image
**Output:** Linked edges

1: Create $\rho$ and $\theta$ for all possible lines
2: Create an array $A$ indexed by $\rho$ and $\theta$
3: **for** for each point $(x, y)$ **do**
4:    **for** Each angle $\theta$ **do**
5:       $\rho = x\cos\theta + y\sin\theta$
6:       $A[\rho, \theta] = A[\rho, \theta] + 1$
7:       **if** $A > Threshold$ **then**
8:          return line
9:       **end if**
10:    **end for**
11: **end for**

---

**FIGURE 1.19**: Green color lines are the lines linked by Hough transform (see ebook for color)

three parallel field lines in bird's eye view is used to indicate the occurrence of the gate, as shown in Figure 1.19.

### 1.4.4 Thresholding

Thresholding is the most basic approach of image segmentation, where the given image is partitioned into regions based on intensity values. Suppose that the gray-level histogram corresponds to an image, $f(x, y)$, composed of white objects in a dark background, in such a way that object and background pixels have gray levels grouped into two dominant modes. One obvious way to extract the objects from the background is to select a threshold $T$ that separates the objects from the background. As shown in Figure 1.20(b), the valley point of the histogram of the cameraman is selected as threshold $T$. Then any point $(x, y)$ for which $f(x, y) > T$ is called an object point; otherwise, the point is called a background point. In other words, the segmented image $g(x, y)$ is given by

$$g(x, y) = \begin{cases} 1 & if \quad f(x, y) > T \\ 0 & if \quad f(x, y) <= T \end{cases}$$

Figure 1.20(c) shows the segmented cameraman image using threshold $T$.

#### 1.4.4.1 Multiple Thresholding

Multiple thresholding classifies a point $(x, y)$ as belonging to the background if $f(x, y) <= T_1$, to one object class if $T_1 < f(x, y) <= T_2$, and to the other object class if $f(x, y) > T_2$. The segmented image $g(x, y)$ is given as

$$g(x, y) = \begin{cases} a & if \quad f(x, y) > T_2 \\ b & if \quad T_1 < f(x, y) <= T_2 \\ c & if \quad f(x, y) <= T_1 \end{cases}$$

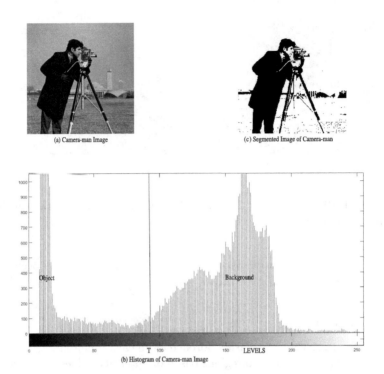

(a) Camera-man Image

(c) Segmented Image of Camera-man

(b) Histogram of Camera-man Image

**FIGURE 1.20**: Segmentation of cameraman image based on threshold $T$

where a,b,c are any three distinct intensity values. The success of thresholding is directly related to the width and depth of the valleys separating the histogram modes. Key factors affecting properties of valleys are: the separation between peaks, the noise content in the image, the relative size of objects and background, the uniformity of the illumination source, and the uniformity of the reflectance properties of the image.

Equation 1.26 represents the threshold selection. If $T$ depends on $f(x,y)$ only, then the selected threshold is called the global threshold. If $T$ depends on pixel location $(x,y)$ and local property in a neighbourhood centered at $(x,y)$, the selected threshold is called the local threshold.

$$T = T[x, y, p(x,y), f(x,y)] \tag{1.26}$$

### 1.4.4.2    Global Thresholding

Algorithm 4 is used to obtain threshold $T$ automatically. In general, if $T_0$ is larger, the algorithm will perform fewer iterations. The initial threshold $T$ can be selected as average intensity of the image. Global thresholding uses a fixed threshold for all pixels in the image. Hence, it works only if the intensity histogram of the given image contains two separate peaks corresponding to

(a)  (b)  (c)

**FIGURE 1.21**: (a) Original image, (b) Global thresholding, (c) Local (adaptive) thresholding

---

**Algorithm 4** Global Thresholding Algorithm

**Input:** Input image
**Output:** Segmented image

Select an initial estimate for $T$
Select a parameter $T_0$ to control the number of iterations
Loop:
Segment the image using $T$ into two groups of pixels: $G_1$ consisting of all pixels with gray level $> T$ and $G_2$ consisting of pixels with values $<= T$.
Compute the average intensity values $m_1$ and $m_2$ for the pixels in regions $G_1$ and $G_2$.
Compute $T_{diff} = T - [(m_1 + m_2)/2]$
Compute new threshold value $T = (m_1 + m_2)/2$
**if** $T_{diff} < T_0$ **then**
    go to Loop
**end if**

---

the objects and background. As shown in Figure 1.21(b), global thresholding failed to segment the river and mountain on the backgorund of the image. Hence, to achieve better segmentaion performance, a local thresholding technique is used.

### 1.4.4.3    Local Thresholding

This approach is based on the assumption that smaller image regions are more likely to have approximately uniform illumination. For each pixel, a threshold is calculated. If the pixel value is below the threshold, it is set to the background value, otherwise it assumes the foreground value. To calcualte local threshold, the local neighbourhood of each pixel is examined statistically. The statistical properties include the mean of the local intensity distribution, the median value, or the mean of the minimum and maximum values. The size of the neighborhood is also important. It has to be large enough to cover sufficient foreground and background pixels. On the other hand, choosing regions that are too large can violate the assumption of approximately uniform illumination. The algorithm that uses adaptive thresholding to achieve better segmentaion is called the adaptive thresholding algorithm. Figure 1.21(c) shows the result of such adaptive thresholding. The threshold will be called optimum if it gives best separation between classes in terms of their intensity values.

## 1.4.5    Region Growing

In noisy images, edge detection is sometimes difficult. In such cases, the region-based approach is preferred. A region is a group of pixels with similar properties. Region growing is the simplest image segmentation approach, that groups pixels or sub-regions into larger regions based on some pre-defined criteria. Initially, one set of seed points are selected. Based on some criteria, all neighbors of seed points are tested. If neighbors are similar to the seed point based on the criteria, neighbors will be merged into the region. This procedure is repeated until the growth of the region is stopped. The similarity measure can be selected as gray level, texture, color, or shape. The selection of seed points, and the similarity criteria, are based on the problem under consideration.

## 1.4.6    Region Splitting and Merging

This approach is based on the homogeneity of the image. First the given image is divided into four parts, and the homogenity of all parts is checked. The parts that are not homogeneous are subdivided into four parts again and homogeneity will be checked. The division will be continued until we

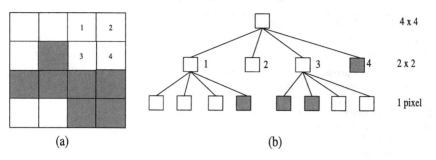

**FIGURE 1.22**: (a) Partitioned image, (b) Quadtree

get all parts homogeneous. Hence, this approach is called region splitting. Also, if the homogeneous parts are similar, then they are merged. Hence, this algorithm is called region splitting and merging. The same thing can be explained considering quadtrees. A quadtree means that each node of trees has four descendants and the root of the tree corresponds to the entire image. Also each node represents the subdivision of a node into four descendant nodes. The subdivision is carried out if the node of the tree does not represent homogeneity. As shown in Figure 1.22, nodes representing region 1 and region 3 are subdivided further. The basics of splitting and merging are discussed in the algorithm.

Advantages:

The image could be split progressively as per the required resolution, since the number of splitting levels is decided by the user.

The homogeneity criteria and merging criteria will be decided by the user.

Disadvantages:

Because of the division into four parts, blocky segments may be produced. This problem can be reduced by increasing the number of splitting levels. But this will increase computation. Hence, trade-offs are decided by the user.

## 1.4.7 Watershed-Based Segmentation

Watershed-based segmentation often produces more stable segmentation. Watershed segmentation is a way of automatically separating particles that touch. The gradient magnitude of an image is considered as a topographic surface, and watershed lines are extracted using gradient magnitude. If the pixel has highest gradient magnitude intensity, it is considered to be a pixel of the watershed line. Watershed lines represents region boundaries and catchment basins represents the objects. Hence, the key behind watershed is to convert the image into another image whose catchment basins are the objects to be segmented. We can find the catchment basins and watershed lines for any grayscale image using Algorithm 5.

---

**Algorithm 5** Watershed Algorithm

---

**Input:** Input image
**Output:** Segmented image

 1: Start with all pixels with lowest possible value
 2: **for** Each intensity level k **do**
 3:     **for** Each group of pixels of intensity k **do**
 4:         **if** adjacent to exactly one existing region **then**
 5:             add these pixels to that region
 6:         **end if**
 7:         **if** adjacent to more than one existing regions **then**
 8:             mark as boundary
 9:         **end if**
10:         **if** not adjacent to any existing regions **then**
11:             start a new region
12:         **end if**
13:     **end for**
14: **end for**

---

### 1.4.7.1    Use of Markers

Because of noise, the watershed segmentation algorithm often leads to over-segmentation of an image. Hence, a concept of markers is used as a solution. Markers limit the number of allowable regions and bring additional knowledge into the segmentation procedure. A marker is a connected component belonging to an image. Markers will be selected based on certain criteria. Markers are classified as external markers and internal markers. External markers are associated with the background and internal markers are associated with the objects of interest. For example, internal markers can be regions surrounded by the higher altitude points, every region should be a connected component, and every point in the region should have same gray-level value. External markers can be some regions of particular background color. Figure 1.23 shows the segmentation of the pears image using the watershed algorithm with markers.

---

## 1.5    Applications

Digital image processing has wide applications. Some of its applications areas are discussed in this section.

### 1.5.1    Television Signal Processing

The image processing techniques are used in Television signal processing to improve image brightness, contrast, and color hue adjustment. Figure 1.24

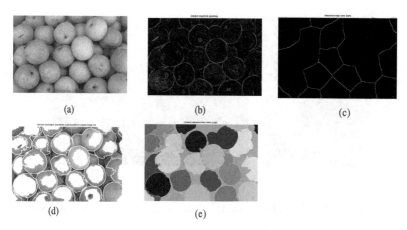

**FIGURE 1.23**: (a) Image of pears, (b) Image gradient, (c) Watershed lines, (d) Image showing markers, (e) Result of segmentation

**FIGURE 1.24**: Adaptive brightness and contrast enhancement

shows the examples of adaptive brightness and contrast enhancement of a television image.

## 1.5.2 Satellite Image Processing

Satellite image processing is used to extract information about natural resources, such as vegetation, forest land, hyderological, mineral, and geological resources. Figure 1.25 shows satellite images giving information about the floods in Bangladesh and geographical information about Bangladesh and northeast India.

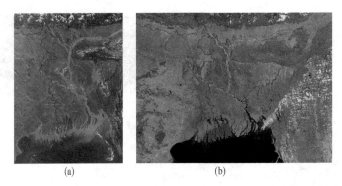

(a)                                    (b)

**FIGURE 1.25**: (a) Floods in Bangladesh, (b) Bangladesh and Northeast India

### 1.5.3    Medical Image Processing

Nowadays, for disease dignosis purposes, imaging devices such as X-rays, computer aided tomographic (CT) images, and ultrasound are used extensively [33]. Examples of medical images acquired by different image modalities such as ultrsound, CT, magnetic resonance imaging (MRI), positron emission tomography (PET), and X-rays are shown in Figure 1.26. The field of medical image processing has played an important role in the development of intelligent healthcare systems, and devices such as image reconstruction and modelling techniques allow instant processing of 2-D signals to create 3-D images in the case of CT-scans. Clinicians and physiologists are always overloaded with the database, which makes it very difficult to communicate proper information. In such cases, machine learning helps to communicate extracted knowledge from the database via user interface. In addition, using advanced image processing techniques, experts can access quantitative anatomical structural changes caused by injuries or other disorders in the brain. Image processing techniques assist in predicting the outcomes and optimal strategies for treatment [11]. From the above examples, we can observe that advancement in technology facilitates health care. The researchers and clinicians must obtain updated advances in image processing techniques for better treatment and diagnosis. It is true that medical image processing has contributed immensely to improving the health and quality of human life [29].

### 1.5.4    Robot Control

Image processing is used for automatic inspection and automatic vehicle driving for robots. First, camera components take the images, and then hardware components generate the appropriate motion and position control signals for robot control.

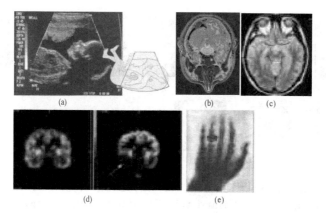

**FIGURE 1.26**: (a) Ultrasound, (b) CT, (c) MRI, (d) PET, (e) x-ray

**FIGURE 1.27**: Videoconferencing showing focus groups or meetings

## 1.5.5 Visual Communications

A videophone is capable of both audio and video duplex transmission. Its purpose is to serve individuals, not groups. However, multipoint videoconferencing allows participants to sit in a virtual conference room and communicate as if they were sitting right next to each other, as shown in Figure 1.27. Until the mid 90s, the cost of hardware required for videoconferencing was expensive to buy. But the situation has changed now. Recent developments in codec technology are increasing the use of relatively low-bandwidth ISDN for various display formats. Since the cost of video conferencing has been reduced, videoconferencing will be one of the fastest growing technologies in the coming years.

## 1.5.6 Law Enforcement

Biometric-based image processing systems will identify criminal suspects accurately, saving time and resources. As shown in Figure 1.28, the biometrics used are ear, face, facial thermogram, hand thermogram, hand vein, hand

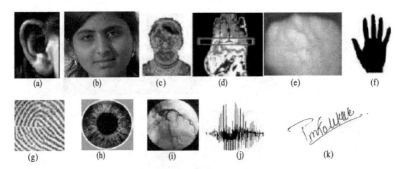

**FIGURE 1.28**: Biometrics: (a) Ear, (b) Face, (c) Facial thermogram, (d) Hand thermogram, (e) Hand vein, (f) Hand geometry, (g) Fingerprint, (h) Iris, (i) Retina, (j) Voice, (k) Signature

geometry, fingerprint, iris, retina, voice, and signature. Law enforcement gets help from these biometric security features.

## 1.6    Summary

In this chapter, various image processing techniques are discussed. The elements of general digital image processing systems are explained in brief. Image segmentation methods are discussed in more detail because of their extensive use in similarity searches of video surveillance systems. Gradient-based edge-detection algorithms such as the Prewitt filter and Sobel filter are very sensitive to noise. The size of such kernel filters and coefficients is fixed and cannot be adapted to a given image. In order to distinguish between valid image contents and noise, a robust algorithm that will adapt the varying noise level is needed. The Canny edge detection algorithm deals with this problem. Though this algorithm is computationally more expensive as compared to Sobel, Prewitt, and Roberts operator, its performance is better under almost all scenarios. The watershed algorithm is another powerful algorithm used to segment an image. Digital image processing has many applications. Its applications areas include medical image analysis, television, security, image compression, and satellite image processing. In this chapter, elementary notions in image processing are presented, but there is still a lot more to explore. If you are new to the image processing field, I hope this chapter will have given you basic information and motivation to study image processing further.

# Chapter 2

## Basics of Video Compression and Motion Analysis

---

## 2.1 Video Compression

### 2.1.1 What Is Video Compression?

Compression reduces the amount of data for transmission and storage purposes. In video compression techniques, redundant video data is removed in order to reduce the size of a digital video file. As a result of that, the space required for storage is reduced and such files can be easily sent over the network. The compression techniques can be classified as data compression and image compression.

Data compression reduces the number of bits required to store or transmit data, by exploiting statistical properties of the data. The data may be in the form of numeric, text, binary, image, or sound. Image compression is based on the fact that there is a strong correlation among pixels or some wider parts of the image. This correlation is exploited to compress the image file. The transform techniques are used for the compression of continuous tone (grayscale) image data. The discrete cosine transform (DCT) has been used for continuous-tone compression.

### 2.1.2 Why Video Compression?

Video compression is required since the size of an uncompressed video file is very large. To help you understand this, the following example shows that an uncompressed video produces a large amount of data. The bandwidth required for transmitting a video with a frame resolution of 720 × 576 pixels (PAL), a refresh rate of 25 fps, and 8-bit color depth is calculated as follows:

$$bandwidth = luminance\ component \times frame\ rate \times color\ depth +$$
$$2 \times (chrominance\ component \times frame\ rate \times color\ depth)$$
$$= 720 \times 576 \times 25 \times 8 + 2 \times (360 \times 576 \times 25 \times 8) = 1.66\ Mb/s \quad (2.1)$$

For high-definition television (HDTV), taking frame rate as 60:

$$1920 \times 1080 \times 60 \times 8 + 2 \times (960 \times 1080 \times 60 \times 8) = 1.99 \ Gb/s \qquad (2.2)$$

Even with powerful computer systems, such large size data requires extremely powerful computational systems for handling the data. Hence, to reduce the size of a digital video file, compression techniques are employed by exploiting the redundancy available in the uncompressed video file. This helps to store the file in less space and transmit with limited bandwidth.

### 2.1.3     Types of Video Compression

#### 2.1.3.1     Lossless

Lossless compression reduces the size of data without the loss of any information, and when reconstructed, restores it to its original form. GIF is an example of lossless images compression, which is generally used for text or spreadsheet files, where losing words or financial data could create a problem. For example, if the picture contains 100 pixels of blue color, it can be compressed by saying 100 *blue pixels* instead of saying 100 times *blue pixel*. Huffman-coding and run-length encoding exploit redundancies and allow high compression ratios without loss of any data.

(a)                              (b)

**FIGURE 2.1**: (a) Lossless, (b) Lossy

### 2.1.3.2 Lossy

Lossy compression removes data based on visual sensory characteristics and limits. When played, the video looks like the original video, but the compressed file is different from the original source. In this approach, the exact original information cannot be obtained by decompression. The example of lossy and lossless results is shown in Figure 2.1. The amount of data reduction possible by using lossy compression is often much higher than what may be done through lossless techniques. The lossy compression technique is based on limitation of viewers. A video file can be compressed until it is not noticable by viewers. Even when noticeable by the user, further data reduction may be desirable based on the application and availablity of the memory space and transmission speed. Higher compression is achieved at the cost of loss of some data.

## 2.1.4 Latency

Compression is achieved by implementing algorithms that remove some image data. When the video is to be viewed, these compression algorithms are applied to interpret the data and view it on the display. Execution of these algorithms will take some time. The delay that occurrs because of the compression algorithm is called compression latency. In video compression, several adjacent frames are being compared, which introduces more latency, as shown in Figure 2.2. Since applications of video compression for TV broadcast and DVD playback do not require real-time interaction, the latency between the source and the decoded video is not important and can easily extend to several seconds. However, in the applications where there is a need for real-time interaction, such as video conferencing and videophone, latency is the most crucial specification of the system as it determines whether the system will be stable or not. Video coding standards such as H.264 and MPEG-4 have achieved very low latency of 8 ms to 4 ms. For some applications such as movies, compression latency is not important, since the video is not watched

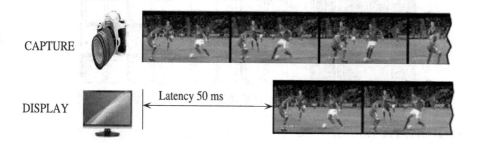

**FIGURE 2.2**: Video broadcasting is delayed by 50 miliseconds due to latency

live. However, for surveillance applications, very low latency is expected for live monitoring of activities especially using PTZ and dome cameras.

## 2.1.5    MPEG Compression

Figure 2.3 shows the steps of the MPEG compression algorithm. A reduction of the resolution is done by using subsampling of chrominance channels. Then motion compensation is performed in order to reduce temporal redundancy. Then discrete cosine transform is computed and its values are quantized. Next, entropy coding using run-length encoding and the Huffman coding algorithm is performed to generate compressed MPEG video.

### 2.1.5.1    Reduction of the Resolution

The human eye has a lower sensitivity to color information than to dark bright contrasts. An RGB color file is converted into a YUV color file to achieve compression by subsampling chrominance channels U and V. The chrominance channels are grouped together based on the type of subsampling (2 or 4 pixel values). The subsampling reduces the data volume by 50% and 33% for the 4:2:0 and 4:2:2 subsampling, respectively, as shown in Figure 2.4. The terms

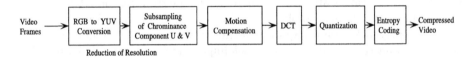

**FIGURE 2.3**: MPEG compression block diagram

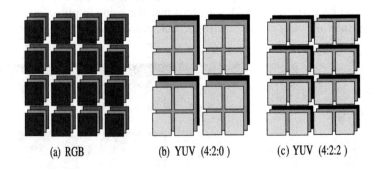

**FIGURE 2.4**: (a) RGB color space (no grouping or subsampling), (b) 4 pixel values of the chrominance channel are grouped together, (c) 2 pixel values of the chrominance channel are grouped together

$YUV$ and $Y'C_bC_r$ refers to an analog encoding scheme and digital encoding scheme, respectively. The only difference between these schemes is that the scale factors on the chroma components are different. However, the term $YUV$ is often erroneously refered as $Y'C_bC_r$.

### 2.1.5.2 Motion Estimation

Often, two successive frames of a video sequence have small differences, except in moving objects. As shown in Figure 2.5, MPEG-standard offers three types of frames in order to reduce this temporal redundancy: 1) Intra-coded frames (I-frames): These are intra predicted and self-contained frames. 2) Predicted frames (P-frames): These frames are predicted from the last I or P reference frame. 3) Bidirectional frames (B-frames): These are predicted from two references, one in the past and one in the future frame. Hence, out-of-order decoding is required for MPEG video decoding purposes.

The I-frames are called intracoded frames, in which frame compression is performed by exploiting spatial redundancies. Their compression is not that high as compared to P and B frames. P and B frames are referred to as inter-coded frames. The P-frames are predicted from an earlier I or P frame. Since only differences are stored, p-frames require less space than I frames. Since B-frames are predicted from the previous and next frame, they are called bidirectional frames (B-frames).

The video decoder reconstructs the video by decoding the bit stream frame by frame. Decoding always starts with an I frame, since it can be decoded independently. P and B frames require reference frames for decoding purposes. The references between the different types of frames are realized by a process

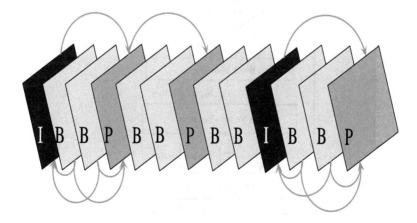

**FIGURE 2.5**: Sequence of frames in an MPEG video separated into I, P, and B frames

called motion estimation or compensation. The correlation between two frames in terms of motion is represented by a motion vector. The motion vectors are usually computed by a motion estimation algorithm based on the pixel value difference in sucessive frames. Higher compression ratios and better quality of the coded video sequence is achieved by good motion estimation. However, motion estimation is not suited for real time applications since it requires large computations. Figure 2.6 shows the flowchart of motion estimation. The steps involved in motion estimation are as follows:

1) Frame Segmentation
The frame is usually divided into non-overlapping macro blocks of size $8 \times 8$ or $16 \times 16$ pixels. The selection of block size is a critical factor in terms of time performance and quality for the following reasons: If block size is small, more vectors need to be calculated. If the blocks are too large, the motion matching is usually less correlated. The block size of $16 \times 16$ pixels is used in

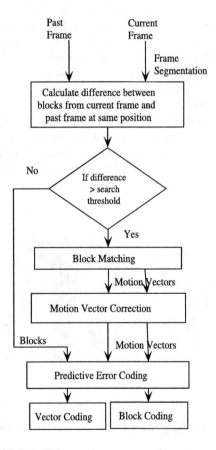

**FIGURE 2.6**: Schematic process of motion estimation

MPEG standards.

## 2) Search Threshold
If the difference between two blocks at the same position is higher than search threshold, the full block is transmitted. In such cases, there is no need to estimate motion.

## 3) Block Matching
Block matching predicts blocks in an actual predicted frame by comparing blocks from previous frames. The block matching is a very time-consuming process. This is because each block of the current frame is compared with the blocks within a search window in the previous frame. Though only luminance information is used to compare the blocks, the color information can be included in the encoding. The selection of search window is a critical factor for the quality of the block matching. If we select the larger search window, it is more likely that we will get a matching block. But large search windows slow down the encoding process drastically. Often rectangular search windows are used since horizontal movement is more likely than vertical movement in any video.

## 4) Prediction Error Coding
Video motions are usually more complex. Predicted motion by shifting of objects in 2-D is not always the same as in the actual scene and this error is called prediction error. For compensating prediction error, the MPEG stream contains a matrix. After prediction, only the difference between the predicted

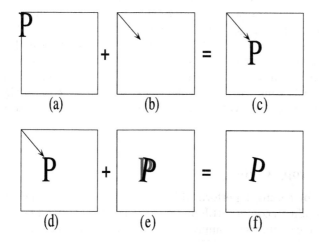

**FIGURE 2.7**: Prediction error coding: Row-1: (a) I-frame, (b) Motion vector, (c) Forward predicted frame, Row-2: (d) Forward predicted frame, (e) Prediction error, (f) Resulting frame

frames and original frames are coded, because less data is required to store only the differences. These differences are shown by red and black regions in Figure 2.7.

5) Vector Coding

After computing the motion vectors and evaluating the correction, the video is compressed. Large parts of MPEG videos consist of B and P frames. Usually B and P frames have mainly stored motion vectors. Therefore, an efficient compression of motion vector data that has high correlation is expected.

6) Block Coding

DCT is used for coding each block, which is explained in the next section.

### 2.1.5.3    Discrete Cosine Transform

DCT is used to represent image data in the frequency domain. The image is divided into blocks of size $8 \times 8$ or $16 \times 16$ pixels. The disadvantages of DCT are computationally very expensive.

### 2.1.5.4    Quantization

During quantization, the DCT terms are divided by the terms in the quantization matrix. This quatization matrix is designed based on human visual perception. Human eyes are more sensitive to low frequencies than high frequencies. Hence, in the quantization matrix, the values of quantization coefficient $Q(u, v)$ are selected such that after quantization, higher frequencies end up with a zero value and we can discard such values. The quantization values are calculated as $F_{Quantised}(u, v) = F(u, v)/Q(u, v)$, where Q is the quantization matrix of size $N \times N$. The $Q$ defines the compression level and the quality of the image. After quantization, the DC and AC terms are stored separately. Since the correlation between the adjacent blocks is high, only the differences between the DC-terms are stored. The AC-terms are stored in a zig-zag path with increasing frequency values. Higher frequency values of AC-terms will be discarded.

### 2.1.5.5    Entropy Coding

The entropy coding is performed in two steps: run length encoding (RLE) coding and Huffman coding. RLE and Huffman coding are lossless compression methods that can further compress data by an additional factor of 3 to 4 depending on the redundancy. The difference between two adjacent frames is generated and DCT is applied to the macroblock of the difference image. Later on it is quantized, and RLE and Huffman coding algorithms are applied to generate a final compressed image. We will see in the next section the different compression formats.

## 2.1.6 Video Compression Standards

The two major groups working on video coding standards are: MPEG (Moving Picture Experts Group) and ITU-VCEG (International Telecommunication Union- Video Coding Experts Group). The MPEG working group was established in 1988 and was formed by the ISO (International Organization for Standardization) and IEC (International Electrotechnical Commission) to set standards for audio and video compression and transmission. The ISO/IEC developed international standards for video compression, and some of their standardized MPEG compression formats are: MPEG-1, MPEG-2, MPEG-3, MPEG-4, MPEG-7, and MPEG-21.

ITU-VCEG is the name of the visual coding part of working party 3 (media coding) of study group 16, which is responsible for multimedia coding, systems and applications of the ITU-T. It is responsible for standardization of the $H.26X$ line of video coding standards. The organization VCEG has standardized the following video compression formats: H.120, H.261, H.263, and H.263V2. These standards are used for video conferencing purposes [3].

Three other video codings have been developed with the collaboration of ISO/IEC (also known as MPEG) with ITU-T and they are: H.262, H.264, and H.265. As shown in Figure 2.8, these groups came up with many differ-

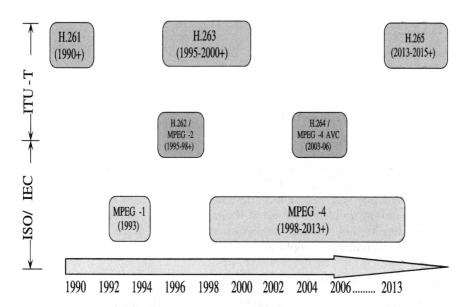

**FIGURE 2.8**: Timeline of release of various video coding standards

ent types of formats for video coding, and all these formats served different applications in their respective time.

**TABLE 2.1**: Comparison Table of Video Compression Standards

| Video Coding Standard | First release year | Publisher | Bit Rate | Applications |
|---|---|---|---|---|
| H.261 | 1988 | ITU | 64 Kb/s | Video conference over ISDN |
| MPEG-1 | 1993 | ISO/IEC | 1.5 Mb/s | Video on digital storage media (CD-ROM) |
| MPEG-2/ H.262 | 1995 | ITU, ISO/IEC | > 2 Mb/s | Digital television, video broadcasting |
| H.263 | 1995 | ITU | <33.6 Kb/s | Video conference, Video telephony, Web video content, video on mobile phone (3GP) |
| MPEG-4 AVC/ H.264 | 2003 | ITU ISO/IEC | 10 to 100 Kb/s | HDTV, BlueRay, improved video compression |
| HEVC/ H.265 | 2013 | ITU, ISO/IEC | 10 to 100 Kb/s | 8K UHD TV, improved video compression |

## 2.2    Motion Segmentation

### 2.2.1    Introduction

Motion segmentation decomposes a video into moving objects and background, as shown in Figure 2.9. This is the first fundamental step of many video surveillance systems. A large number of researchers have focused on the segmentation problem because of its importance in surveillance systems. However, despite these efforts, the performance of most of the algorithms still falls far behind human perception.

#### 2.2.1.1    Issues in Motion Segmentations

In an automatic motion segmentation algorithm, the important issues are as follows:

1) Noise:

The motion segmentation performance degrades because of noise. In the case of underwater motion segmentation, there are some specific subsea phenomena such as water turbidity, marine snow, rapid light attenuation, strong reflections, back-scattering, non-uniform lighting, and dynamic lighting. These subsea phenomena degrade the underwater image quality drastically [43].
2) Blurring of data:
When motion is involved, blurring may occur, and because of blurred object boundaries, the performance of motion segmentation degrades.
3) Occlusion:
Sometimes moving object get occluded fully or partially. Sometimes the full object may disappear for some duration and reappear again in the scene. Under such situations, the motion segmentation performance also degrades.

### 2.2.1.2    Main Attributes of a Motion Segmentation Algorithm

Object segmentation based on motion is a very important step of any video surveillance system. It is not always possible to get prior knowledge about the shape of the object and number of objects in the scene [59]. Under such situations, accuracy of the extraction of the object based on motion segmentation is very important. The main attributes of a motion segmentation algorithm are as follows:
1) Feature-based or dense-based:
In feature-based methods, the objects are extracted based on features such as corner points or salient points, whereas dense methods compute a pixel-wise motion [64].
2) Occlusions:
A motion segmentation algorithm should be capable of handling partial or full occlusion cases.
3) Multiple objects:
A motion segmentation algorithm should be capable of dealing with multiple objects in the scene.

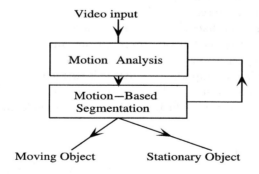

**FIGURE 2.9**: Flowchart of motion segmentation algorithm

4) Spatial continuity:
This is the ability to exploit spatial continuity.
5) Temporary stopping:
Motion segmentation should handle temporary stoppage of the objects in the scene.
6) Robustness:
A motion segmentation algorithm should extract objects even with noisy images in the video sequences. For underwater videos, robustness of the algorithm is very important since underwater images are usually noisy.
7) Missing data:
Motion segmentation algorithms should handle the cases of missing data.
8) Non-rigid object:
Motion segmentation algorithms should deal with with non-rigid objects.

### 2.2.2    Motion Segmentation Algorithms

Motion segmentation algorithms can be divided into the following categories:
1) Image Difference
2) Statistical Theory Based
3) Optical Flow
4) Layers
5) Factorization

#### 2.2.2.1    Image Difference

Image difference is one of the easiest and most widely used techniques for detecting moving objects. Despite its simplicity, this technique cannot be used as it is because it is very sensitive to noise.

In this approach, the image difference of successive frames is taken and the thresholding technique is applied to get moving objects in the scene. This gives a coarse map of the temporal changes. This technique provides good results when the camera is fixed. If the camera is moving, the complete image is changing and extracting moving objects is difficult. In order to deal with such situations, the rough map of the intensity-changing areas is computed, and for each blob, spatial or temporal information is extracted. This approach can be modified to handle noise and illumination changes. This will be discussed in detail in the next chapter, which is on background modelling.

#### 2.2.2.2    Statistical Theory

Since in motion segmentation each pixel has to be classified as background or foreground, it can also be seen as a classification problem. Statistical approaches such as maximum a posteriori probability (MAP), particle filter (PF) and expectation maximization (EM) provide a general tool to extract the moving objects. MAP is based on the Bayes rule, which is mentioned in Equation 2.3.

$$P(\omega_j|x) = \frac{p(x|\omega_j)p(\omega_j)}{\sum_{i=1}^{C} p(x|\omega_i)p(\omega_i)}, \quad (2.3)$$

where $x$ is the pixel, $\omega_1...\omega_C$ are the $C$ classes (usually background or foreground), $P(\omega_j|x)$ is the a posteriori probability, $p(x|\omega j)$ is the conditional density, $P(\omega_j)$ is the apriori probability and $\sum p(x|\omega_i)p(\omega_i)$ is the density function. MAP classifies $x$ as belonging to the class $\omega$ which maximizes the a posteriori probability. MAP is often used in combination with other techniques. In [52] MAP is combined with a probabilistic data association filter, and in [7], MAP is combined with level sets to extract foreground objects.

### 2.2.2.3 Optical Flow

Optical flow (OF) is a vector that describes the distribution of the apparent velocities of brightness patterns in the video sequence. It was first proposed in 1980 by Horn and Schunck for image sequence [19]. But the idea of using discontinuities in the OF for motion segmentation is even older. Lucas and Kanade [48] also proposed a method to compute OF. The main limitations of OF-based motion segmentation is that OF is very sensitivite to noise, and also, the number of computations required are very large. Nowadays, because of high-speed computers and improvements in the algorithms, OF is widely used.

### 2.2.2.4 Layers

This technique is based on depth information. Its main purpose is to understand different depth layers, and which objects lie in which layer. This approach is often used in stereo vision problems. The depth information of the objects also helps in handling occlusion problems. In [43] a method is proposed for representing the scene into layers without computing the depth. In this approach, first the coarse-moving components between every pair of frames are computed. Then the image is divided into patches, and the rigid transformation that moved the patch from one frame to the next is located. The initial estimate is then refined using $\alpha$-$\beta$ swap and $\alpha$-expansion [19] minimization algorithms.

### 2.2.2.5    Factorization Technique

Tomasi and Kanade [58] proposed a factorization technique to extract the moving objects using features tracked through a sequence of images. Factorization methods are widely used because of their simplicity. In this approach, first the trajectory matrix T containing the position of F features tracked throughout N frames is computed. Then matrix T is factorized into motion M and structure S matrices. If the origin of the world coordinate system is moved at the centroid of all the feature points, and in the absence of noise, the matrix T is at most rank 3. Exploiting this constraint, T can be decomposed and truncated using a singular value decomposition technique. Because the rows of the motion matrix are orthonormal, the matrix M can be evaluated using orthogonality constraints, and finally decomposed, up to a scale factor. Though this method gives the 3-D structure of the object and the motion of the camera, it is very sensitive to noise and unable to deal with missing data and outliers. However, from this initial structure-from-motion approach, many approaches have been proposed for motion segmentation.

## 2.3    Optical Flow Methods

Motion detection is the process of detecting a change in the position of an object relative to its background or a change in the background relative to an object. Optical flow is the distribution of apparent velocities of movement of brightness patterns in an image. It can arise from relative motion of objects and the camera. Optical flow gives a description of motion and thus is used in motion detection. Optical flow is very useful to study motion while the object is stationary and the camera is moving, while the object is moving and the camera is stationary, or when both are moving. Optical flow is based on two assumptions: 1) The brightness of the object point is constant over time. 2) Points that are nearby in the frame move in a similar manner. Since video is simply a sequence of frames, the optical flow methods calculate the motion of an object between two frames that are taken at times $t_1$ and $t_2$, as shown in Figure 2.10.

### 2.3.1    Estimation of Optical Flow

Optical flow methods are based on local Taylor series approximations of the image signals, i.e., they use partial derivatives with respect to the special and temporal coordinates. Suppose a voxel at location $(x, y, t)$ with intensity $f(x, y, t)$ moves by $dx$, $dy$, and $dt$ between two frames, then by brightness constancy assumption, we can write

$$f(x, y, t) = f(x + dx, y + dy, t + dt) \qquad (2.4)$$

Assuming the movement to be small and applying Taylor series in the above equation, we get

$$f(x + dx, y + dy, t + dt) = f(x, y, t) + \frac{\partial f}{\partial x} dx + \frac{\partial f}{\partial y} dy + \frac{\partial f}{\partial t} dt \qquad (2.5)$$

From Equations 2.4 and 2.5, we can write

$$f_x dx + f_y dy + f_t dt = 0 \qquad (2.6)$$

where $f_x$, $f_y$, and $f_t$ are the derivatives of the image at $(x, y, t)$ in the corresponding directions.

$$f_x u + f_y v + f_t = 0 \qquad (2.7)$$

where $u = \frac{dx}{dt}$ and $v = \frac{dy}{dt}$ are the x and y components of the velocity or optical flow of $f(x, y, t)$. Equation 2.7 above represents one equation with two unknown variables $u$ and $v$ (the aperture problem). To estimate these unknown variables $u$ and $v$, several methods are proposed. The two most widely used methods are: 1) Horn-Schunck optical flow estimation, and 2) Lucas-Kanade optical flow estimation.

### 2.3.1.1  Horn-Schunck Optical Flow Estimation

Optical flow equation 2.7 can be written as straight line form as follows:

$$v = -\frac{f_x}{f_y} - \frac{f_t}{f_y} \qquad (2.8)$$

Figure 2.11 shows the line represented by Equation 2.8. The flow perpendicular to this line from the origin is called normal flow and denoted by $d$. The value of $d$ is calculated as

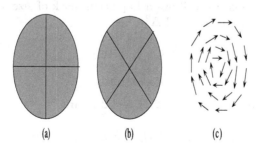

**FIGURE 2.10**: (a) at time $t_1$, (b) at time $t_2$, (c) Optical flow

$$d = \frac{f_t}{\sqrt{f_x^2 + f_y^2}} \tag{2.9}$$

The normal flow $d$ is fixed. However, parallel flow $p$ changes, which is why we have to estimate it. Horn and Schunck considered this problem as an optimization problem with two constraints: 1) Brightness constancy, 2) Smoothness constraint. The optimization function used is as follows:

$$\iint \{(f_x u + f_y v + f_t)^2 + \lambda(u_x^2 + u_y^2 + v_x^2 + v_y^2)\} dx dy \tag{2.10}$$

where $\lambda$ expresses relative effect of the smoothness constraint. We can use variational calculus to solve this.

Differentiating Equation 2.10 w. r. t. u, we will get

$$(f_x u + f_y v + f_t)f_x + \lambda(\Delta^2 u) = 0 \tag{2.11}$$

Differentiating Equation 2.10 w. r. t. v, we will get

$$(f_x u + f_y v + f_t)f_y + \lambda(\Delta^2 v) = 0 \tag{2.12}$$

where

$$\Delta^2 u = \frac{\partial^2 u}{\partial x^2} + \frac{\partial^2 u}{\partial y^2} \tag{2.13}$$

and

$$\Delta^2 v = \frac{\partial^2 v}{\partial x^2} + \frac{\partial^2 v}{\partial y^2} \tag{2.14}$$

$$\Delta^2 u = u_{xx} + u_{yy} \tag{2.15}$$

$$\Delta^2 v = v_{xx} + v_{yy} \tag{2.16}$$

In the discrete version, we will use a Laplacian mask of size $4 \times 4$, shown in Figure 2.12, to compute $\Delta^2 u$ and $\Delta^2 v$.

$$\Delta^2 u = u - u_{avg} \tag{2.17}$$

$$\Delta^2 v = v - v_{avg} \tag{2.18}$$

The discrete version of Equations 2.11 and 2.12 are

$$(f_x u + f_y v + f_t)f_x + \lambda(u - u_{avg}) = 0 \tag{2.19}$$

Differentiating Equation 2.10 w. r. t. v, we will get

$$(f_x u + f_y v + f_t)f_y + \lambda(v - v_{avg}) = 0 \tag{2.20}$$

Solving the above Equations 2.19 and 2.20

$$u = u_{uvy} - f_x \frac{P}{D} \tag{2.21}$$

$$v = v_{avg} - f_y \frac{P}{D} \tag{2.22}$$

where

$$P = f_x u_{avg} + f_y v_{avg} + f_t \tag{2.23}$$

$$D = \lambda + f_x^2 + f_y^2 \tag{2.24}$$

Algorithm 6 presents Horn-Schunck optical flow estimation technique. Horn-Schunck's method is more accurate but requires a large number of iterations.

### 2.3.1.2 Lucas Kanade Optical Flow Estimation

Lucas and Kanade considered optical flow in a neighborhood close to the pixels. He considered a window size of $3 \times 3$ pixels and computed optical flow for all 9 pixels.

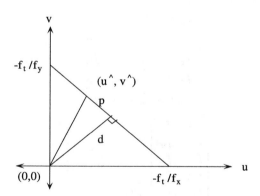

**FIGURE 2.11**: Optical flow estimation

| 0 | -1/4 | 0 |
|---|------|---|
| -1/4 | 1 | -1/4 |
| 0 | -1/4 | 0 |

**FIGURE 2.12**: Laplacian mask

---

**Algorithm 6** Horn-Schunck Optical Flow Estimation Method

---

**Input:** Recorded video frames
**Output:** Estimated motion vector

1: Initialize $k = 0$
2: Initialize $u^k, v^k$
3: Compute $u$ and $v$ values using the following equations

$$u^k = u_{avg} - f_x \frac{P}{D}$$

$$v^k = v_{avg} - f_y \frac{P}{D}$$

where $P = f_x u_{avg} + f_y v_{avg} + f_t$ and $D = \lambda + f_x^2 + f_y^2$
4: $k = k + 1$
5: Repeat until some error measure is satisfied

---

$$
\begin{aligned}
f_{x1}u + f_{y1}v &= -f_{t1} \\
f_{x2}u + f_{y2}v &= -f_{t2} \\
f_{x3}u + f_{y3}v &= -f_{t3} \\
f_{x4}u + f_{y4}v &= -f_{t4} \\
f_{x5}u + f_{y5}v &= -f_{t5} \\
f_{x6}u + f_{y6}v &= -f_{t6} \\
f_{x7}u + f_{y7}v &= -f_{t7} \\
f_{x8}u + f_{y8}v &= -f_{t8} \\
f_{x9}u + f_{y9}v &= -f_{t9}
\end{aligned}
\tag{2.25}
$$

$$
\begin{bmatrix}
f_{x1} & f_{y1} \\
f_{x2} & f_{y2} \\
f_{x3} & f_{y3} \\
f_{x4} & f_{y4} \\
f_{x5} & f_{y5} \\
f_{x6} & f_{y6} \\
f_{x7} & f_{y7} \\
f_{x8} & f_{y8} \\
f_{x9} & f_{y9}
\end{bmatrix}
\begin{bmatrix} u \\ v \end{bmatrix}
=
\begin{bmatrix}
-f_{t1} \\
-f_{t2} \\
-f_{t3} \\
-f_{t4} \\
-f_{t5} \\
-f_{t6} \\
-f_{t7} \\
-f_{t8} \\
-f_{t9}
\end{bmatrix}
$$

The above equation can also be written as

$$A\mu = f_t \tag{2.26}$$

Since A is not a square matrix, we cannot invert it, so we will multiply both sides of the above equation by $A^T$

$$A^T A\mu = A^T f_t \tag{2.27}$$

Hence, motion vector $\mu$ is calculated as

$$\mu = (A^T A)^{-1} A^T f_t \tag{2.28}$$

Least Squares Fit Approach:

Another way to get the estimation of motion vectors is to minimize squares of optical flow, i.e.,

$$min \sum_i (f_{xi}u + f_{yi}v + f_{ti})^2 \tag{2.29}$$

Differentiating w. r. t. $u$, we get

$$\sum_i (f_{xi}u + f_{yi}v + f_{ti})f_{xi} = 0 \tag{2.30}$$

Differentiating w. r. t. $v$, we get

$$\sum_i (f_{xi}u + f_{yi}v + f_{ti})f_{yi} = 0 \tag{2.31}$$

Equation 2.30 can be simplified as

$$\sum_i f_{xi}^2 u + \sum_i f_{xi}f_{yi}v = -\sum_i f_{xi}f_{ti} \tag{2.32}$$

Equation 2.31 can be simplified as

$$\sum_i f_{xi}f_{yi}u + \sum_i f_{yi}^2 v = -\sum_i f_{yi}f_{ti} \tag{2.33}$$

Equation 2.32 and 2.33 can be written in matix form as

$$\begin{bmatrix} \sum_i f_{xi}^2 & \sum_i f_{xi}f_{yi} \\ \sum_i f_{xi}f_{yi} & \sum_i f_{yi}^2 \end{bmatrix} \begin{bmatrix} u \\ v \end{bmatrix} = \begin{bmatrix} -\sum_i f_{xi}f_{ti} \\ -\sum_i f_{yi}f_{ti} \end{bmatrix}$$

$$\begin{bmatrix} u \\ v \end{bmatrix} = \begin{bmatrix} \sum_i f_{xi}^2 & \sum_i f_{xi}f_{yi} \\ \sum_i f_{xi}f_{yi} & \sum_i f_{yi}^2 \end{bmatrix}^{-1} \begin{bmatrix} -\sum_i f_{xi}f_{ti} \\ -\sum_i f_{yi}f_{ti} \end{bmatrix}$$

$$\begin{bmatrix} u \\ v \end{bmatrix} = \frac{1}{\sum_i f_{xi}^2 \sum_i f_{yi}^2 - (\sum_i f_{xi}f_{yi})^2} \begin{bmatrix} \sum_i f_{yi}^2 & -\sum_i f_{xi}f_{yi} \\ -\sum_i f_{xi}f_{yi} & \sum_i f_{xi}^2 \end{bmatrix} \begin{bmatrix} -\sum_i f_{xi}f_{ti} \\ -\sum_i f_{yi}f_{ti} \end{bmatrix}$$

The Lucas Kanade method works better for small motion. If an object moves faster, a $3 \times 3$ mask fails to estimate spatiotemporal derivatives. Hence, pyramids can be used to compute optical flow vector.

## 2.4    Applications

Digital video processing has many applications. In this section some important video processing applications are discussed in brief.

### 2.4.1    Surveillance and Security

Research in the surveillance system area is growing continuously and substantially, because of the security-related incidents such as terrorist acts in many countries around the world. This results in the need for intelligent surveillance, and monitoring systems, consisting of real-time image capture, transmission, processing, and surveillance information understanding. This information will be vital to people's safety, and indeed to national security. In this book, we will discuss many intelligent video surveillance systems in detail.

### 2.4.2    Content-Based Video Indexing and Retrieval

Nowadays, because of developments in technology, capturing video is very easy. As a result of that, very large video databses and libraries are generated. Automatic video indexing and retrieval tools are required to browse the videos from large databses. The key challenge in the area of media management is automation of content annotation, indexing [28], retrieval, summarization, search, and browsing applications. One of the major challenges is reducing the semantic gap between the simplicity of features that can be computed automatically and the richness of semantics in user queries posed for video search and retrieval. Content-based video indexing for sports [36], news [37], and movies have been developed.

**FIGURE 2.13**: Video event sequence for exciting clips of $Goal_B$ and $Redcard_A$

### 2.4.3   Automatic Highlight Generation of Sports Videos

Since sports videos are globally widespread, automatic sports video analysis has drawn increasing attention. As video capturing, processing, and transmitting technology has developed rapidly, it is much easier to record and archive digital video data. This has increased the need necessities for automatic video analysis of sports video sequences. Since sports videos are well structured in nature, researchers have proposed many techniques for automatic sports video analysis. However, the broadcasting styles of different sports vary drastically. So, researchers have targeted the individual games, such as soccer [34], tennis, cricket [38] [39], basketball, baseball, volleyball, and golf.

In [40], an automatic highlight-generation technique for a soccer sports TV broadcast is discussed. The proposed system first detects exciting clips based on audio features [34] and then classifies the individual scenes within the clip into events such as *replay, player, referee, spectator, and players gathering*. The higher-level semantic concept-labels are assigned to the exciting clips using a Bayesian belief network-based approach [28]. The higher level concept-labels are *goals, saves, yellow-cards, red-cards*, and *kicks* in soccer videos. Figure 2.13 shows the results of higher-level concepts label assignment to soccer video clips based on lower-level event-labels [40]. The labeled clips are selected based on their degree of importance for generating soccer highlights.

### 2.4.4   Traffic Monitoring

Video surveillance for monitoring traffic is becoming an increasingly popular field of application. Traffic cameras are placed above traffic signals on

**FIGURE 2.14**: Traffic monitoring using a surveillance system

road crossings and intersections along busy roads. They are becoming more popular day by day, as they help traffic police to levy fines on the lawbreakers. As shown in Figure 2.14, this type of surveillance system will automatically identify persons breaking the traffic rules and inform the traffic police. The camera footage is extremely helpful for the transportation department, for future development of roads based on the camera footage of traffic patterns and accident-prone zones.

---

## 2.5    Summary

In this world where technology is improving exponentially day by day, it is getting difficult to store or broadcast large amounts of huge video files. To solve this issue, groups such as ITU-T and MPEG are standardizing video compression formats that are capable of compressing video files by tenfolds. This enables users to store video files on personal computers or mobile phones with limited storage, and it also enables streaming video files in less time. In this chapter, various video compression techniques and major video compression formats are discussed in detail. The optical flow-based motion estimation approach is disussed in detail, since it is used for a variety of low-level tasks like motion parameter, depth, or segmentation. These tasks are often used as input for many higher-level tasks, like autonomous navigation, object tracking, image stabilization, video compression, and human activity recognition. Applications of digital video processing for surveillance, indexing and retrieval, automatic sports highlight generation, and traffic monitoring are discussed in brief to get an idea of the implementation of video processing techniques.

# Chapter 3

# Background Modeling

## 3.1 What Is Background Modeling?

Background modeling is a technique used in computer vision for the detection of foreground objects in a frame sequence. The accuracy of foreground object detection depends on the accuracy in the modeling of the background. In order to detect moving objects, first a background model of the scene is constructed. Then each video frame is compared against a background model. The pixels that deviate significantly from the background are considered to be pixels of moving objects. After background subtraction, we get a raw foreground image containing only moving object. Preprocessing techniques such as erosion and dilation operations are applied to remove the noise. As shown in Figure 3.1, if the difference between the current frame and the background frame is greater than the threshold value, then the foreground object is detected; otherwise no foreground object is detected. After getting the foreground object, the centroid of the object is computed and the object can be tracked with reference to its centroid in the scene. The background model should be robust against environmental changes such as changes in illumination, and able to identify all moving objects in the scene. The general steps of background subtraction-based object detection and tracking architecture are shown in Figure 3.2.

## 3.2 Background Modeling Techniques

Moving objects can be easily detected if we remove the background from the scene. A simple approach for detecting moving objects in video sequences is background subtraction from the current frame. Hence, initially the background model must be constructed without the moving objects. Then the current frame is subtracted from the background model to extract moving objects. This technique is used in the surveillance applications for object tracking, traffic monitoring, and video compression. Background modeling tecniques can be broadly classified based on non-statistical and statistical approaches.

In the non-statistical approaches, the first frame is considered to be the background and the subsequent frames are subtracted from the background. Then the pixels with a value higher than a threshold are considered to be the objects. Each pixel in a frame is considered either a part of the moving object or the background. The background is updated along the frame sequences. Since non-statistical approaches are fast, they are suitable for real-time appli-

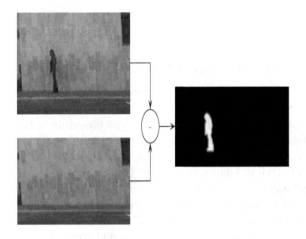

**FIGURE 3.1**: Background extraction

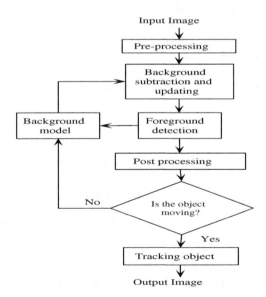

**FIGURE 3.2**: Flow-chart for background extraction

cations.

In the statistical approaches, the probability distribution functions of the background pixels are estimated. Then in each video frame, the likelihood that a pixel belongs to the background is computed. The statistical-based approaches have better performance compared to the non-statistical-based approaches in modeling background of the outdoor scenes. However, they require more memory and processing time. These methods are slower than the non-statistical methods. The Gaussian mixture model (GMM)-based background modeling is one of the popular techniques of statistical background modeling. In this approach, mixture of models is used to represent the statistics of the pixels in the scene. The multimodal background modeling is very useful in removing repetitive motion such as shining water, leaves on a branch, a fluttering flag.

## 3.2.1 Non-Statistical Background Modeling Methods

In this approach, the initial frame is considered to be the background. The background is updated along the frame sequence. In this approach, moving objects are detected by using the difference between the current frame and the background model. Since fewer number of computations are required for background extraction, it is suitable for real-time applications. In order to detect moving objects, subsequent frames are subtracted from the background and then pixels with the intensity value higher than a threshold are considered to be the pixels of moving objects.

### 3.2.1.1 Background Modeling Independent of Time

Since this method is the simplest approach for computing background that is independent of time, it is called background modeling independent of time (BMIT). In this approach, the first frame in video sequences is considered to be the background and remains unchanged along the video sequences. The mathematical description of the background model can be represented as

$$B_{x,y}^k = I_{x,y}^0 \tag{3.1}$$

where $I_{x,y}^k$ is the pixel (x,y) of the $k^{th}$ captured frame, and $B_{x,y}^k$ is the pixel (x,y) of the $k^{th}$ background model. In cases of complex background or dynamic background, where illumination levels and background objects change abruptly, this method fails as it considers only the first frame to be the background.

### 3.2.1.2 Improved Basic Background Modeling

Background modeling independent time suffers from noise and varying luminance in image sequence. The improved basic background modeling (IBBM) method was developed to remove the deficiencies of the BMIT approach. Once the pixel value of the absolute difference frame is more than the threshold value, the pixel is regarded as part of the moving objects; otherwise it is assigned to the background. Whenever a pixel belongs to the moving object, it should be updated; otherwise, it is not essential to update. According to this idea, the mathematical description of IBBM can be expressed as follows:

$$B_{x,y}^k = \begin{cases} I_{x,y}^k, & AD_{x,y}^k < T \\ B_{x,y}^{k-1}, & AD_{x,y}^k > T \end{cases} \tag{3.2}$$

where $AD_{x,y}^k$ is the pixel $(x,y)$ of the absolute difference frame between the $k^{th}$ captured frame and the $(k-1)^{th}$ background model as shown below:

$$AD_{x,y}^k = |I_{x,y}^k - B_{x,y}^{k-1}| \tag{3.3}$$

Although the IBBM method reduces the noise effect and the varying luminance effect, it has a deckle effect. Hence, it suffers from updating the deckle of the foreground in the background model.

### 3.2.1.3  Long-Term Average Background Modeling

The Long Term Average Background Model (LTABM) was developed to overcome the deckle-effect problem of IBBM. LTABM is defined by the following equation:

$$B_{x,y}^k = \frac{1}{k} \sum_{r=1}^{k} I_{x,y}^k \tag{3.4}$$

It is recursive in nature and is given by

$$B_{x,y}^k = (1 - \frac{1}{k})B_{x,y}^{k-1} + \frac{1}{k}I_{x,y}^k \tag{3.5}$$

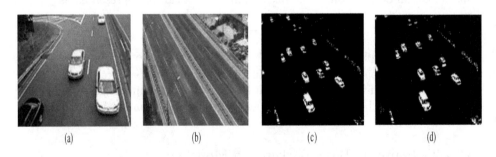

(a)                    (b)                    (c)                    (d)

**FIGURE 3.3**: (a) Current frame, (b) Background frame, (c) Extracted foreground image with threshold k=10, and (d) Extracted foreground image with threshold k=50. (See ebook for color.)

From the above equations, it is clear that LTABM is actually related to the frame averaging technique. So, when the number of frames is quite large, the weights of each frame are too small. But when the number of frames is less, the weights of each frame are large. Figure 3.3 presents the estimated background and extracted foreground by changing the value of $k$. Figure 3.3 (c) and (d) present the extracted foreground using estimated background with $k = 10$ and $k = 50$, respectively.

## 3.2.2 Statistical Background Modeling Methods

Statistical background modeling techniques take the first few frames as the training sample, and by calculating the probability distribution function, they estimate the pixels that correspond to background. Hence, these pixel locations are used as background location to model a uniform background of a video stream. These approaches are widely used in dynamic backgrounds, outdoor situations, and real-time scenarios. The statistical background modeling method based on GMM [56] is widely used, and hence it is explained in the next subsections. It is useful for modeling data that comes from one of several groups. The groups might be different from each other, but data points within the same group can be well-modeled by a Gaussian distribution.

### 3.2.2.1 Example of GMM

Suppose the price of a randomly chosen blue ball is normally distributed with mean \$8.00 and standard deviation \$1.50. Similarly, the price of a randomly chosen orange ball is normally distributed with mean \$12 and variance \$2.00. Here, the price of a randomly chosen ball is not normally distributed. This is made very clear by looking at the fundamental property of the normal distribution. It is highest near the center, and quickly drops off as you get farther away. But, the distribution of a randomly chosen ball is bimodal. As shown in Figure 3.4, the center of the distribution is near \$10, but the probability of finding a ball near that price is lower than the probability of finding a ball for a few dollars more or a few dollars less. Still, it is not normally distributed. It is too wide and flat in the center. GMM is used to model the distributions where we have several groups and the data within each group is normally distributed.

### 3.2.2.2 GMM Model

There are two types of clustering methods:
1) Hard clustering:
In this approach, clusters do not overlap and the elements either belong to a cluster or not. For example: K-means clustering.

2) Soft clustering:

In soft clustering, clusters may overlap. For example: Mixture models.
Mixture models are a probabilistic way of doing soft clustering in which each cluster is assumed to be a generative model whose parameters are unknown. One such mixture model is the GMM, which is used for background modelling. This background modeling is very useful in removing repetitive motion, such as a fluttering flag and leaves on a branch. In GMM, we assume that in a given frame, each pixel is generated by combining multiple multivariate Gaussians instead of a single Gaussian. For $d$ dimensions, the Gaussian distribution of a vector $x = (x^1, x^2, , x^d)$ is given by

$$N(x|\mu, \Sigma) = \frac{1}{(2\pi|\Sigma|)^{\frac{1}{2}}} exp\{\frac{-1}{2}(x - \mu)^T \Sigma^{-1}(x - \mu)\} \qquad (3.6)$$

where $\mu$ is the mean and $\Sigma$ is the covariance of the Gaussian. Now each pixel in the scene can be modelled as a linear super-position of $K$ Gaussians given by the probability

$$p(x) = \sum_{k=1}^{K} \pi_k N(x|\mu, \Sigma) \qquad (3.7)$$

where $\pi_k$ is the prior probability or mixing coefficient of the $k^{th}$ Gaussian and $K$ is the total number of Gaussians such that $\sum_{k=1}^{K} \pi_k = 1$ and $0 \leq \pi_k \leq 1$. Expectation maximization is one of the most popular approaches for assignment of parameters, i.e., mean, covariances, and mixing coefficients to the given mixture model $\theta = \{\pi_1, \mu_1, \Sigma_1, ...\pi_k, \mu_k, \Sigma_k\}$.

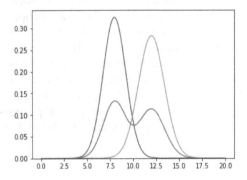

**FIGURE 3.4**: Probability density function for blue ball (blue color). Probability density function for orange ball (orange color). Probability density function for both balls (green color). (See ebook for color.)

### 3.2.2.3 Expectation Maximization GMM Algorithm

Problem statement:

Given a set of data $X = \{x_1, x_2, ......., x_N\}$ drawn from a distribution of GMM, estimate the parameters $\theta = \{\pi_1, \mu_1, \Sigma_1, ...\pi_k, \mu_k, \Sigma_k\}$ of the GMM model that fits the data where $N$ is total number of data points.

Solution:

Maximize the likelihood $p(X|\theta)$ of the data with regard to the model parameter $\theta$ comprising the means, covariances, and the mixing coefficients.

$$\theta^* = argmax \, p(X|\theta) = argmax \prod_{i=1}^{N} p(x_i|\theta) \tag{3.8}$$

Steps of EM algorithm:

Step 1:

Initialize $\pi_k, \mu_k, \Sigma_k$ and evaluate the initial value of log likelihood.

Step 2 (expectation (E) step):

Evaluate the responsibilities or posterior probabilities using the current parameter values with the help of the Bayes rule:

$$\gamma_k = p(k|x) = \frac{p(k)p(x|\theta)}{p(x)} \tag{3.9}$$

$$\gamma_k = \frac{\pi_k N(x|\mu, \Sigma)}{\sum_{k=1}^{K} \pi_k N(x|\mu, \Sigma)} \tag{3.10}$$

where $\gamma_k$ is the responsibility or the latent or hidden variable.

Step 3 (maximization (M) step): Re-estimate or update the parameters using the current responsibilities $\gamma_k$, i.e., $\pi_k, \mu_k, \Sigma_k$ are the parameters to be estimated.

$$\mu_k = \frac{\sum_{n=1}^{N} \gamma_k(x_n)x_n}{\sum_{n=1}^{N} \gamma_k(x_n)} \tag{3.11}$$

$$\Sigma_k = \frac{\sum_{n=1}^{N} \gamma_k(x_n)(x_n - \mu_k)^T \Sigma^{-1}(x_n - \mu_k)}{\sum_{n=1}^{N} \gamma_k(x_n)} \tag{3.12}$$

$$\pi_k = \frac{1}{N} \sum_{n=1}^{N} \gamma_k(x_n) \tag{3.13}$$

Step 4 (evaluate log likelihood):

$$ln \, p(x|\mu, \Sigma, \pi) = \sum_{n=1}^{N} \{ \sum_{k=1}^{K} \pi_k N(x|\mu_k, \Sigma_k) \} \tag{3.14}$$

Here $p(x|\mu, \Sigma, \pi)$ is the log likelihood function. Find the maximum likelihood calculated that truly represent the data samples. If there is no convergence, return to Step 2.

---

**Algorithm 7** EM-GMM Algorithm

---

**Input:** Given a set of data $X = \{x_1, x_2, ....., x_N\}$
**Output:** Maximum likelihood model $\theta$

1: Initialize mean $\mu_k$, covariance matrix $\Sigma_k$, and mixing coefficient $\pi_k$
2: **for** iteration $t$ **do**
3:    **for** each value $k$ **do**
4:       Estimate $\gamma_k$ using Equation 3.10 (E-step)
5:       Find $\mu_k$, $\Sigma_k$ and $\pi_k$ (M-step)
6:    **end for**// $K$
7: **end for**// $T$ iterations
8: **for** data $n$ **do**
9:    **for** each value $k$ **do**
10:       Find log likelihood function value
11:    **end for**// $K$
12: **end for**// $N$ Data points
13: Find $argmax\ (ln\ p(x|\mu, \Sigma, \pi))$
14: **if** $\theta = \theta_{t-1}$ **then**
15:    Return $\theta_t$
16: **else**
17:    go to step 2
18: **end if**

---

#### 3.2.2.4    GMM-Based Background Detection

Background detection can be performed using GMM-based Algorithm 7. A particular pixel in the given dataset is regarded as background if probability distribution, i.e., $p(x)$ for that pixel, is greater than a particular threshold, i.e., $p(x) > T$ where $T$ is the threshold. Generally, a high-value probability density is assigned to pixels belonging to the background. GMM-based background detection works well but we have to guess the number of Gaussians. Kernel density estimation can be used for that type of guessing. The final result depends on the initial choice of starting points. The convergence can be ensured when change in $p(x_1, x_2, ..x_N)$ is sufficiently small. Figure 3.5 shows

(a)                                (b)

**FIGURE 3.5:** (a) Traffic scene, (b) Extracted background using GMM

the result of GMM-based background detection.

Advantages:
1) A different threshold is selected for each pixel.
2) These pixel-wise thresholds are adapting by time.
3) Objects are allowed to become part of the background without destroying the existing background model.
4) It provides fast recovery.

Limitations:
1) Cannot deal with sudden, drastic lighting changes
2) Initializing the Gaussians is important.
3) There are relatively many parameters, and they should be selected intelligently.

## 3.3 Shadow Detection and Removal

The performance of many video surveillance algorithms, such as object segmentation, object detection, object classification, and object tracking de-

(a)       (b)

**FIGURE 3.6**: (a) Short shadow during afternoon, (b) Long shadow during evening

grade because of shadows of various objects in the scene. Since shadows of objects also move along with the objects, adjacent objects may get connected through shadows, and object recognition systems will get confused. Hence, shadow detection and removal is one of the important preprocessing steps. Decomposing a moving object and its shadow is a difficult problem because of illumination changes. Many techniques have been proposed over the years, but shadow detection still remains an extremely challenging problem when dealing with outdoor scenes. Shadows occur when objects occlude light from a light source. As shown in Figure 3.6, shadows provide rich information about object shapes as well as light orientations. In outdoor environments, we can guess the time of video recording based on length of shadows. The short shadow as seen in Figure 3.6 (a) indicates that video was recording during the afternoon, and the long shadow shown in Figure 3.6 (b) indicates that video was recording in the evening. But shadows still need to be removed because they may cause errors in visual applications such as video object classification and tracking. Moreover, if objects have intensities similar to those of shadows, shadow removal could become extremely difficult.

Foreground detection is one of the most important and critical fields in video surveillance. Improvements in background modeling have led to numerous applications, such as event detection, object behavior analysis, suspicious object detection, and traffic monitoring. Dynamic backgrounds pose a problem because they treat swaying trees and moving shadows as part of the foreground. So an algorithm for detection of these moving shadows and their removal is necessary to prevent misclassification in foreground detection. Studies on influences of moving shadows classify them into four models, i.e., color model, statistical mode, textural model, and geometric model [65].

1) Color Model: This uses the differences between shaded and non-shaded pixels.

2) Statistical Model: It uses the probablistic functions to determine whether the pixel belongs to the shadow or not.

3) Texture Model: It uses the property that the texture of the foreground object is completely different from that of the background, and all the textures are uniformly distributed inside a shaded region.

4) Geometric Model: It uses the geometric properties of the object to detect the foreground and shadow.

The color model attempts to describe the color change of the shaded pixel and find the color feature that is illumination invariant. Black and white cameras rather than color cameras would be much more popular for outdoor applications. The shadow removal method based on the color model might not work in such situations. The principle behind the texture model is that the texture of the foreground is different from that of the background, while the texture of the shaded area remains the same as the background. Similarly, the texture model could have better results under the unstably illuminating conditions without the color information. If the objects in the scene are texture-less, texture models may give poor performance [26]. The geometric model could be more adaptive to the specific scenes due to its dependency upon the geometric relations between objects and scenes, and so far it has been mainly applied in simulated environments. Because of large computations, geometric models are not suitable for real-time cases. A method used for shadow removal in traffic videos is discussed in detail in the next section.

## 3.3.1 Shadow Removal for Traffic Flow Detection

In this section, the method proposed by C. T. Lin et al. [44] is discussed: a fast-moving shadow removal scheme that combines texture and statistical models. This method proved to be stable and used the texture model instead of the color model to simplify the system. Furthermore, statistical methods are used to improve the performance of the system, to successfully deal with textureless objects. This method was initially proposed for removing shadows of moving vehicles caused by nonuniform distributions of light reflections in the daytime in real traffic situations. In this approach, GMM is used for background modeling and removal of moving shadows for a practical application of traffic flow detection and vehicle counting in intelligent transportation systems. The shadow removal is based on edge and gray level combined features as shown in Figure 3.7. Algorithm 8 and Figure 3.8 provide more implementation details. The above technique successfully detected the real objects and neutralized the negative influences of shadows. Compared with other previous approaches, the method proposed in [44] can accurately detect the foreground objects without shadows.

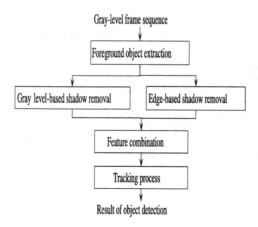

**FIGURE 3.7**: Flow chart of shadow removal technique

---

**Algorithm 8** Moving Shadow Removal Algorithm

---

**Input:** Video Sequence
**Output:** Foreground images without shadow

1: **Background Modeling**
2: Apply GMM-based approach to obtain background and foreground object.
3: **Edge-Based Shadow Removal**
4: Extract edges using Sobel operator on background image to get $BI_{Edge}(x,y)$
5: Extract edges using Sobel operator on foreground image to get $FO_{Edge_{MBR}}(x,y)$
6: Pixel by Pixel Maximization
7: $MI_{edge_{MBR}}(x,y) = max(FO_{edge_{MBR}}(x,y), BI_{edge}(x,y))$
8: $St_{edge_{MBR}}(x,y) = MI_{edge_{MBR}}(x,y) - BI_{edge}(x,y)$
9: Adaptive Binarization
10: Calculate $t_{final}(x,y)$ at location (x,y)
11: $t_{final}(x,y) = m(x,y)[1 + k(\frac{s(x,y)}{R}) - 1] + Th_{supress}$
12: Where m(x,y) and s(x,y) are the mean and standard deviation of mask centered at pixel $(x,y)$, respectively. $Th_{supress}$ is supression term whose value is set to 50 empirically.
13: $BinI_{MBR}(x,y) = \begin{cases} 0, & if \ \ St_{edge_{MBR}}(x,y) \le t_{final}(x,y) \\ 255 & otherwise \end{cases}$
14: Boundary Elimination
15: A $7 \times 7$ mask is applied to achieve boundary elimination. If region covered by mask completely belongs to the foreground objects, reserve the point, otherwise eliminate the point.
16: **Gray-Based Shadow Removal**
17: Gaussian Darkening Factor Model Updating
18: Shadow potential pixels are selected and darkening factor is calculated for each gray level; the model is updated and thus we get the final Gaussian darkening factor model.
19: Determination of Non-Shadowed Pixel
20: Difference between mean of Gaussian model and darkening factor is calculated and if the difference is smaller than 3 times the standard deviation, it is classified as shadowed pixel, otherwise non-shadowed pixel.
21: **Feature Combination**
22: Combine the two feature images by applying OR operation on $Ft_{Edgebased_{MBR}}$ and $Ft_{DarkeningFactor_{MBR}}$. Then apply filtering and dilation operations on feature integrated image, and thus foreground without shadow is obtained.

---

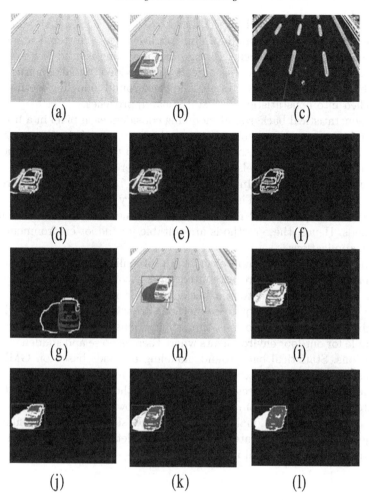

**FIGURE 3.8**: Output image of different steps of moving shadow removal algorithm 3.7: (a) Background image, (b) Foreground object, (c) $BI_{edge}$, (d) $FO_{edge_{MBR}}$, (e) $MI_{edge_{MBR}}$, (f) $BinI_{MBR}$, (g) $Ft_{edgebased_{MBR}}$, (h) Red points selected for Gaussian model, (i) $Ft_{Edgebased_{MBR}}$, (j) $Ft_{DarkeningFactor_{MBR}}$, (k) Feature integrated image, (l) Foreground image after dilation

## 3.4 Summary

In the past, real-time video applications were not possible because of computational limitations of the system. But nowadays, very high-speed DSP processors are available to carry out video processing application, and this overcomes the computational barrier faced by the conventional systems. Systems should be robust to handle illumination changes, cluttered areas, and overlapping regions in the scene. Background modeling is used to extract the region of interest such as moving objects in the video, by subtracting background frame from the picture frame. The background modeling methods can be classified into nonstatistical and statistical approaches.

The nonstatistical background modeling considers each pixel in a frame to be either as part of the moving object or the background. In this approach, the first frame is considered to be background and the subsequent frames are subtracted from it. The pixels having value higher than the threshold value are considered the objects. In this approach, the background gets refreshed along the frame sequences. These methods are very easy to implement. Also, these methods require small memory and less time as compared to statistical methods. Hence, these methods are suitable for indoor environments and real-time applications.

In the statistical-based approach, the probability distribution functions of the background pixels are estimated, and accordingly, the likelihood of whether the pixels belong to the background or not is checked. However, memory and time required for these methods are very high. Hence, statistical approaches are not suitable for real-time applications. However, these methods are suitable for outdoor environments where there is noise and sudden changes in conditions. Statistical background modeling methods based on GMM are widely used and hence are explained in the detail in this chapter.

In the moving object detection process, one of the main challenges is to differentiate moving objects from the shadows. Shadows are usually misclassified as parts of moving objects, making analysis stages such as object classification and object tracking inaccurate. Hence, shadow detection and removal techniques are discussed in detail in this chapter.

# Part II

# Object Tracking

# Chapter 4

## Object Classification

Object classification detects moving objects in a video sequence and classifies them into categories such as humans, vehicles, birds, clouds, or animals. Although the problem of classifying moving objects is very trivial for human beings, it is very challenging for a computer algorithm to do the same with human-level accuracy. Also, the algorithm should be invariant to the number of variations in order to successfully identify and classify moving objects. For example different illumination conditions, different scale and viewpoint variations, deformations and occlusions may degrade the performance of the algorithm.

First foreground objects are detected, and then they are classfiied into the different classes and subclasses. Figure 4.1 shows objects in the video frames classified as person, dog, horse, car, and motorbike. Classification of moving foreground objects is mainly done with shape-based classification or motion-based classification. Shape-based classification uses the amount of dispersion of the foreground object, area, and apparent aspect ratio as key features to classify elements into a single human, group of humans, vehicle, or any other moving object. Some of the human actions use the periodicity property, which can be used to classify humans from other moving objects in the scene.

**FIGURE 4.1**: Objects classified as person, dog, horse, car, and motorbike. (See ebook for color.)

## 4.1    Shape-Based Object Classification

Shape-based features for classifying the object is one of the major challenges, because of the changes in illumination, scale, pose, and camera position in the scene. In automated surveillance, the moving objects in the foreground region such as points, boxes, silhouettes, and blobs are used for classification. Lipton et al. [47] have classified moving objects into human, vehicles, and noise using the blob dispersedness feature. Dispersedness of an object is calculated as follows:

$$Dispersedness = \frac{Perimeter^2}{Area} \tag{4.1}$$

Since human body shape is complex in nature, humans will have more dispersedness than vehicles. Hence, humans can be classified from vehicles using dispersedness. The selection of proper features is very important in obtaining accurate classification.

## 4.2    Motion-Based Object Classification

Motion-based object classification is a very important step in visual surveillance systems [20]. First, silhouette of foreground objects is extracted. Then the features based on the Euclidean distance between the centroid and the boundary pixel of the silhouette are extracted. These shape-based features are compared with the templates corresponding to various actions stored in the database to classify the object. The human body is non-rigid and articulated. Human body motion during any activity is usually periodic in nature. For example, activities such as walking, clapping, and running are periodic in nature. Hence, this property of human motion can be used for classification of humans from other objects in the video frame. Cutler et al. [8] tracked objects of interest and its self-similarity is computed over time. For a periodic motion, computed self-similarity is also periodic.

### 4.2.1    Approaches

First, the object is detected and segmented from video using the background subtraction technique. In silhouette template-based classification, the foreground object distance signal is computed and its similarity with the stored templates is found using the minimum distance. Based on minimum distance, the object is classified into categories such as human, group of humans, vehicle, and animals. The object classification can can be performed online as well as offline.

In the offline case, various objects such as humans, vehicles and animals, their distance signal have computed and stored in the database. In the online case, the silhouette of an object is extracted in each frame and its distance signal is computed. Then computed distance signal is compared with the distance signal of the templates of various classes stored in the database. The object is classified into the the class whose template has the minimum distance from the foreground object.

### 4.2.2 Applications

Motion-based object classification finds huge applications in visual surveillance systems. Usually, in a visual surveillance system, the segmented moving foreground object is humans, vehicles, animals, or birds. Once the class of foreground object is detected, the later tasks such as personal identification, object tracking, and activity recognition can be done much more efficiently and accurately. There are numerous applications for object classification in the fields of robotics, photography, and security. Robots commonly take advantage of object classification and localization in order to recognize certain objects within a scene. Facial recognition techniques are subsets of object recognition that are very useful in applications such as photography and security.

---

## 4.3 Viola Jones Object Detection Framework

In 2001, Paul Viola and Michael Jones proposed an algorithm for real-time object detection. Actually, this algorithm is primarily used for face-detection due to its robust and almost real-time performance. Although it was developed for object detection, it was motivated mainly by the face-detection problem. Also, note that it can only detect faces, and it cannot solve the problem of face-matching. Also, this method requires full-view frontal upright faces, i.e., the entire face must point towards the camera and should not be tilted to either side. This is a very robust method for face detection application, and it successfully distinguishes faces from non-faces.

The Viola Jones algorithm [60] scans a sub-window capable of detecting faces across a given input image. The standard image processing approach would be to rescale the input image to different sizes and then run the fixed size detector through these images. Conversely, Viola Jones rescales the detector instead of the input image and runs the detector through the image many times, and each time with a different size. A Viola Jones detector is a scale-invariant detector that requires the same number of calculations whatever the size. This detector is constructed using an integral image and some simple rectangular features reminiscent of Haar wavelets.

Figure 4.2 presents the steps of the algorithm for visual object detection. This algorithm is capable of processing images extremely rapidly and achieving high detection rates. First, Haar-like features are selected. Then, a new image representation called *Integral Image* is created. This helps to compute the features used by the detector quickly. The learning algorithm based on AdaBoost selects a small number of critical visual features from a larger set and yields extremely efficient classifiers. The last step is to combine more complex classifiers in a *cascade*. This cascade classifier allows background regions of the image to be quickly discarded while spending more computation on promising object-like regions. The implementation of the face detector is quite straightforward, but the training of cascading takes time. Training is slow but the detection is fast.

### 4.3.1   Haar Features

The Viola-Jones algorithm uses Haar like features, which are scalar products of images and Haar-like templates. In practice, the five rectangular patterns shown in Figure 4.3 are used in face detection techniques. The derived features are assumed to hold all the information necessary to detect faces. Since faces are usually regular by nature, the use of Haar features (rectangular features) is justified. Figure 4.4 shows how to apply rectangular features on a given image. Figure 4.4 (c) indicates features associated with the pattern shown in (b) of the original image shown in (a) using the following equation. We will get more of a response when the pattern is matched with the eyebrow region, as shown in Figure 4.4 (c).

$$Value = \sum(pixels\ in\ white\ area) - \sum(pixels\ in\ black\ area) \qquad (4.2)$$

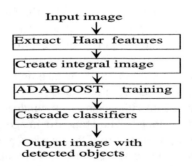

**FIGURE 4.2**: Algorithm for visual object detection

### 4.3.2 Integral Image

The rectangular features are computed easily by converting the given image into an integral image. To calculate the integral image, the sum of all the pixels above and to the left of the pixel concerned is calculated and substituted as the new pixel value. The pixel value of the integral image shown in Figure 4.5 is calculated using the following equation:

$$ii(x, y) = \sum_{x' <= x, y' <= y} i(x', y') \tag{4.3}$$

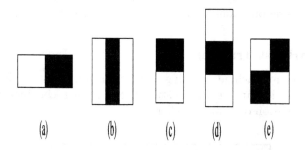

(a)       (b)       (c)    (d)    (e)

**FIGURE 4.3**: Five Haar templates

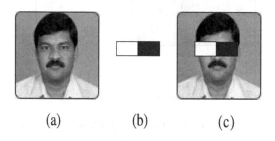

(a)                (b)                (c)

**FIGURE 4.4**: Application of Haar feature on given image

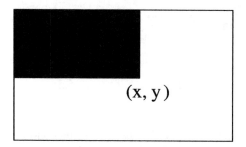

(x, y)

**FIGURE 4.5**: Integral image

where $ii(x, y)$ represents the pixel value of the point $(x, y)$ of the integral image, and $i(x, y)$ represents the pixel value of the point $(x, y)$ in the original image. Figure 4.6 shows the technique of calculating rectangular features. The formula for calculating the sum of pixels in the rectangle $D$ is as follows:

$$sum_D = ii(4) + ii(1) - ii(2) - ii(3) \tag{4.4}$$

where, $ii(1) = A$, $ii(2) = A + B$, $ii(3) = A + C$, and $ii(4) = A + B + C + D$.

$$sum_D = (A + B + C + D) + A - (A + B) - (A + C) = D \tag{4.5}$$

Thus, the integral image allows us to calculate the of sum of all the pixels in any rectangle using only four values at the corner of the rectangle.

### 4.3.3    AdaBoost Training

Adaboost is a machine learning algorithm proposed by Freund and Schapire [13]. It is capable of constructing a strong classifier through a weighted combination of weak classifiers.

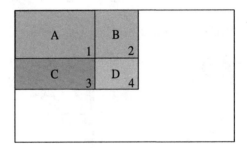

**FIGURE 4.6**: Extraction of rectangular features using integral image

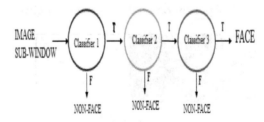

**FIGURE 4.7**: The first stage eliminates 50% of the non-faces, and subsequent classifiers use more scaled directional filters and omit more non-faces

### 4.3.4 Cascading of Classifiers

The basic principle of the Viola-Jones face-detection algorithm is to scan the image using window size 24 × 24 many times through the same image in different sizes. If we apply this algorithm to an image containing only a few faces, a large amount of the evaluated sub-windows will be non-faces. Hence, instead of finding faces, the algorithm should discard non-faces. With the cascaded classifier, at each stage, non-face blocks are discarded and face blocks are considered for the next stages.

As a given sub-window passes through more stages of the cascade, AdaBoost is used to train the classifiers. Starting with a two-feature strong classifier, the face filter is obtained by adjusting the strong classifier threshold from Adaboost to minimize the false negatives. By lowering the threshold, detection rates and false positive rates increase, which explains the situations where false positives exist. Each stage of the cascade is essentially another strong classifier obtained from AdaBoost with a different threshold at each step to maximize the accuracy of the algorithm. Figure 4.7 shows the cascading filter in action.

### 4.3.5 Results and Discussion

Figure 4.8 shows the result of face detection using the Viola-Jones algorithm. This detector is most effective only on frontal images of faces. It can hardly cope with 45-degree face rotation around both the vertical and horizontal axes. It is also possible to get multiple detections of the same face, due to overlapping sub-windows.

---

## 4.4 Object Classification Using Convolutional Neural Networks

In recent times, deep neural networks have been applied to a multitude of problems to achieve very good performance. In particular, convolutional neural networks (CNN) have shown very good results in image classification [42], image segmentation [16], and computer vision problems [15], [57]. In these works, image classification, multi-object localization, and detection are performed using variants of CNNs.

Recent developments in artificial intelligence and deep learning have paved the way to accomplishing tasks involving multimodal learning. Visual questioning-answering is one such challenge that requires high-level scene interpretation from images combined with language modeling of the relevant question and answer. Image-caption generation is the problem of generating a descriptive sentence for an image. A quick glance at an image is sufficient for a human to point out and describe an immense amount of details about the visual scene. The fact that humans can do this with remarkable ease makes this a very interesting and challenging problem for AI, combining aspects of computer vision, in particular, scene understanding. However, this remarkable ability has proven to be an elusive task for our visual recognition models.

### 4.4.1    What Are Convolutional Neural Networks?

CNNs are a special kind of neural networks for processing data having a grid-like topological structure. Examples include 1-D audio waveform data, 2-D image data, and 3-D video data. A typical layer of a convolutional network consists of three stages described in the following sections.

(a)

(b)

**FIGURE 4.8**: Results of face detection after applying the Viola-Jones algorithm

#### 4.4.1.1 Convolution Stage

In the first stage we will use a number of kernels of normally very small dimension, generally $3 \times 3$, $4 \times 4$, or $5 \times 5$, and slide them over the input image to create a feature map. As we slide the kernel over the image, we add up the element-wise dot product of the kernel values and the section of the image it is sliding over. As the same kernel is operated over the image, it is a very memory-efficient operation. The kernels used in a layer are independent of each other, and thus results can be computed extremely quickly in a graphical processing unit (GPU). The convolution operation between a 2-D image I and a 2-D $3 \times 3$ kernel K is given by Equation 4.6.

$$S(i,j) = (K \times I) = \sum_{m=1}^{3} \sum_{n=1}^{3} I(i-m, j-n)K(m,n) \tag{4.6}$$

An example convolution operation between a $4 \times 4$ dimensional image and a $2 \times 2$ dimensional kernel is shown in Figure 4.9.

#### 4.4.1.2 Non-Linear Activation Stage

In the next stage, we apply an element-wise non-linear activation function over the resultant dot product. Mostly *tanh* or *relu* activation is used. The *tanh* operation is given in Equation 4.7.

$$A(i,j) = tanh(S(i,j)) = \frac{sinh(S(i,j))}{cosh(S(i,j))} = \frac{1 - e^{2 \times S(i,j)}}{1 + e^{2 \times S(i,j)}} \tag{4.7}$$

The *relu* operation is given in Equation 4.8.

$$A(i,j) = max(A(i,j), 0) \tag{4.8}$$

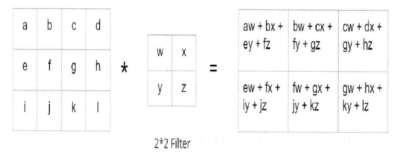

2*2 Filter

Input Image                    Resultant Feature Map

**FIGURE 4.9**: Example of convolution operation

### 4.4.1.3   Pooling Stage

Pooling is a sample-based discretization process where the objective is to down-sample representation to provide an abstracted form of the input. It allows for assumptions to be made about features contained in the sub-regions. Pooling allows later convolutional layers to work on larger sections of the data, because a small patch after the pooling layer corresponds to a much larger patch before it. They also are invariant to some very small transformations of the data. A pooling function is applied to the *tanh* or *relu* activated output at every sliding location, resulting in a summary statistic. Generally, max-pooling or average-pooling is used, which down-samples the image into a smaller dimension. A demonstration is shown Figure 4.10.

If the network is several layers deep, the next convolution operation will be done on the max or average pooled output. After performing the series of convolution, activation, and pooling operations, the output is fed into a fully connected network to classify the image. At an abstract level, CNN can be thought of as a neural network that uses many identical copies of the same neuron which allows the network to have lots of neurons, and express computationally large models while keeping the number of actual parameters, the values describing how neurons behave fairly small. This trick of having multiple copies of the same neuron is roughly analogous to the abstraction of functions in mathematics.

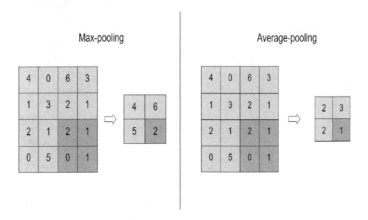

**FIGURE 4.10**: Different types of pooling operations

## 4.4.2   Convolutional Neural Network Models

### 4.4.2.1   Two-Layer Convolutional Neural Network

In this model, two convolutional and max pooling layers are stacked after one another as shown in Figure 4.11 (a). Each convolutional layer has 32 kernels of size $3 \times 3$, and max pooling was performed on every $2 \times 2$ pixels. The output of this was flattened and fed into a multi-layer perceptron network for classification.

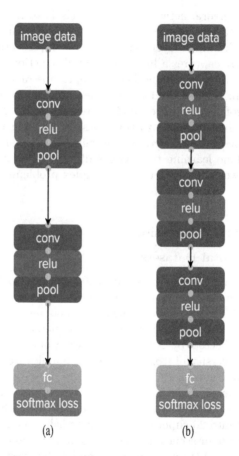

FIGURE 4.11: (a) Two-layer-deep convolutional neural network for classification. (b)Two-layer-deep convolutional neural network for classification. Notations used: conv: convolution layer, relu: activation function, pool: max pooling layer, fc: fully connected layer. (See ebook for color.)

### 4.4.2.2    Three-Layer Convolutional Neural Network

In this model, three convolutional and max pooling layers are stacked after one another as shown in Figure 4.11 (b). Each convolutional layer has 32 kernels of size $3 \times 3$, and max pooling was performed on every $2 \times 2$ pixels. The output of this was flattened and fed into a multi-layer perceptron network for classification.

### 4.4.2.3    Intuition for Using Deep Neural Networks

The objective in neural-network-based models is to find an appropriate representation of data in order to perform a machine learning task. In deep neural networks, each hidden layer maps the input representation to a representation that usually captures a higher level of abstraction. As we go deeper in the network, the representations become more abstract and increasingly more informative. In a supervised machine learning setup, the final output layer is typically a softmax classifier, and the rest of the network learns a representation that is then provided to this classifier, making the classification task easier. This process of discovering or learning effective and useful features is key for any machine learning task. Generally, deeper networks have more representational power and are better for complex problems.

## 4.4.3    Results and Discussion

### 4.4.3.1    Experimental Datasets

1) CIFAR10 Dataset: This consists of 60,000 $32 \times 32$ color images in 10 classes with 6,000 images per class. There are 50,000 training images and 10,000 test images. The classes are airplane, automobile, bird, cat, deer, dog, frog, horse, ship and truck. These classes are completely mutually exclusive. This dataset was collected by Alex Krizhevsky, Vinod Nair, and Geoffrey Hinton from the University of Toronto. A few examples from the CIFAR 10 dataset are shown in Figure 4.12.

2) CIFAR100 Dataset: This is a more detailed description of the CIFAR10 dataset. It consists of 100 classes with 600 images in each class. The 100 classes are grouped into 20 super-classes such as fish, flowers, man-made objects, and natural outdoor scenes. Each of the super-classes contains 5 classes. Overall, there are 50,000 training examples and 10,000 test examples in the CIFAR100 dataset.

**TABLE 4.1**: Test Accuracies for Different Models

| Models | CIFAR10 | CIFAR100 |
|---|---|---|
| SVM | 64.51 | 32.50 |
| Logistic Regression | 63.20 | 27.96 |
| K Nearest Neighbor | 52.49 | 25.23 |
| Two-layer-deep CNN | 73.05 | 51.27 |
| Three-layer-deep CNN | 73.69 | 54.60 |

#### 4.4.3.2 Results and Discussion

In order to show the implementation aspects of the object classification problem, we have developed a machine-learning algorithm using deep CNN, and presented results in the CIFAR10 and CIFAR100 datasets. We observed that two-layer and three-layer simple deep CNN performs better than classifiers such as the support vector machine, logistic regression and nearest neighbor algorithms as shown in Table 4.1. More results on the fashion article classification problem are available in our paper [2].

**FIGURE 4.12**: CIFAR10 dataset. (See ebook for color.)

## 4.5    Object Classification Using Regional Convolutional Neural Networks

A regional convolutional neural network (RCNN) is a state-of-the-art visual object detection system [16] that combines bottom-up region proposals with rich features computed by a convolutional neural network. The goal of the RCNN is to take in an image and correctly identify where the main objects are in the image. This is done using a bounding box method. The input is the image and the outputs are bounding boxes and labels for the object contained in each of the bounding boxes. This is done by first proposing random regions in the image, then extracting features of that region, and then classifying those regions based on their features. In essence, we have turned object detection into an image classification problem. Although RCNN gives very promising results in several multi-object detection datasets, it is a bit slower, as the selective search method is initialized using proposed random regions in the image.

### 4.5.1    Steps of RCNN Algorithm

A region-based convolutional neural network consists of the following three simple steps:
1) Scan the input image for possible objects using an algorithm called Selective Search, generating around 2,000 region proposals. At a high level, selective search looks at the image through windows of different sizes, and for each size, it tries to group together adjacent pixels by texture, color, or intensity to identify objects.
2) Run CNNs on top of each of these region proposals to extract features from that region.
3) Take the output of each CNN and feed it into a) a classifier to classify the region, and b) a linear regressor to tighten the bounding box of the object, if such an object exists. So, to tighten up the boxes, the inputs are sub-regions of the image corresponding to objects and the outputs are new bounding box coordinates for the object in the sub-region.

### 4.5.2    Results and Discussion

Results for the RCNN algorithm are shown in Figure 4.13. The images are from the Microsoft COCO [45] and ImageNet [10] dataset. It is to be observed that the algorithm works very well for all the images. It outputs the bounding box for all the main objects present in the image and classifies the object present in that bounding box with very good accuracy. Overall the results of the RCNN algorithm are very promising.

**FIGURE 4.13**: Object classification results using RCNN. (See ebook for color.)

## 4.6 Summary

In this chapter, object classification techniques are discussed. Shape-based and motion-based objection classifications are two main approaches. Some special approaches such as the Viola-Jones object detection framework for face detection, deep CNN-based object classification, and RCNN-based object classification are discussed in detail. Also, detection of a face and recognizing its emotion are very useful to understanding the behavior of the person in surveillance systems. In order to create interest of the reader in implementation, algorithms are discussed with some results in this chapter.

# Chapter 5

## Human Activity Recognition

Human activity recognition is a popular area of research in the field of computer vision. It is the basis of applications in many areas such as security, surveillance, human computer interaction, and video conferencing. The ability to recognize humans and describe their activities from information acquired by video cameras is a really challenging problem in intelligent video surveillance systems. Difficulties in human activity recognition can arise because of abrupt human motion, changing appearance of a human, changing background of a scene, occlusions, and camera motion. The shape of persons performing the activity can be different. Also, the speed and style of performing any activity varies from person to person.

To date, many approaches have been proposed to recognize simple human activities. However, complex human activity recognition is still a challenging task. The nature of human activities poses a lot of challenges in recognizing concurrent activities, interleaved activities, ambiguity of interpretation, and multiple residents in the scene. Robustness of features and classifiers is required for detecting activity in real time automatically. Features extracted should be scale and rotation invariant as the distance and angle of the person with respect to the camera is usually varying. Motion history image (MHI)-based human activity recgnition using Hu moments is discussed in this chapter in detail. The hidden Markov model (HMM)-based classifier is discussed in detail. Dynamic time warping (DTW)-based activity recognition using the support vector machine (SVM) approach is also discussed in detail.

## 5.1 Motion History Image-Based Human Activity Recognition

### 5.1.1 Motion History Image

Since a video is a sequence of frames, and action cannot be predicted from a single frame, we need a technique to integrate several frames into one to represent motion. For this we use the MHI technique, in which pixel intensities depend upon the history of motion at that location. The brighter values correspond to more recent motion, as shown in Figure 5.1.

The procedure for generating an MHI image is as follows:

1. After the segmentation, the foreground image is obtained, which is a binary image where the body parts in motion have a pixel value of 1. We call this binary function $B(x, y, t)_\tau$.

2. Create an MHI $B(x, y, t)_\tau$ for each action. We need to choose $\tau$ that shows the complete action; best probably is the $t$ where the action just ends.

Given the sequence $B_t$, MHI image M at time t is defined as:

$$M(x, y, t)_\tau = \begin{cases} \tau & if B(x, y)_t - 1 \\ max(M(x, y, t-1)_\tau - 1, 0) & if B(x, y)_t = 0 \end{cases}$$

$\tau$ needs to be chosen such that it captures the full extent of the action. If you use a different $\tau$ for each action, you should scale your MHI image by a value to make the maximum the same for all MHIs. Thus we obtain the MHIs for each of the representative actions as well as the input video sequence.

An MEI is obtained by binarizing the MHI. Since it consists of only two values for intensity, it does not represent the history of motion but shows the

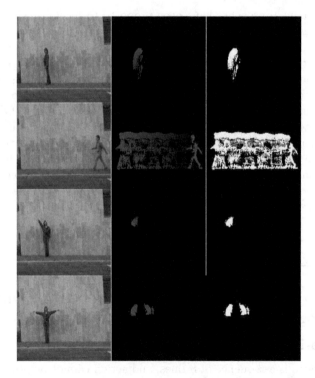

**FIGURE 5.1**: Subjects performing bending, walking, one-hand wave and two-hand wave activity, and their MHI and MEI

spatial location of motion in various frames.

## 5.1.2 Hu Moments

A shape descriptor is a very important descriptor of an object, because the shape of a human changes while performing any action, such as bending. Contour-based shape descriptors and region-based shape descriptors are two important types of shape descriptor. Hu moments are the most popular contour-based shape descriptors. They are a set of moments derived by Hu [23]. These Hu moments recognize the objects irrespective of their position, size, and orientation. The 2-D moments of a digitally sampled image are computed as

$$m_{pq} = \sum_{x=0}^{M-1} \sum_{y=0}^{M-1} x^p . y^q . f(x,y) \qquad p,q = 0,1,2,3,.... \qquad (5.1)$$

where $f(x,y)$ is a digitally sampled image of size $M \times M$.
The moments of image $f(x,y)$ translated by an amount $(a,b)$ is

$$\mu_{pq} = \sum_{x=0}^{M-1} \sum_{y=0}^{M-1} (x+a)^p . (y+b)^q . f(x,y) \qquad (5.2)$$

Now, we can calculate the central moments using Equation 5.1 or 5.2 on substitution of $a = -\bar{x}$ and $b = -\bar{y}$.

$$\bar{x} = \frac{m_{10}}{m_{00}} \quad and \quad \bar{y} = \frac{m_{01}}{m_{00}} \qquad (5.3)$$

$$\mu_{pq} = \sum_{x} \sum_{y} (x - \bar{x})^p . (y - \bar{y})^q . f(x,y) \qquad (5.4)$$

On application of a scaling normalization, central moments become

$$\eta_{pq} = \frac{\mu_{pq}}{\mu_{00}^{\gamma}}, \qquad (5.5)$$

where $\gamma = [(p+q)/2] + 1$. Hu defined seven values that are computed by normalizing the central moments through order three. These moments are invariant to object scaling, position, and orientation. We have the seven mo-

**TABLE 5.1:**    Hu Moments of Images in Figure 5.2

| Hu Moments | $M_1$ | $M_2$ | $M_3$ | $M_4$ | $M_5$ | $M_6$ | $M_7$ |
|---|---|---|---|---|---|---|---|
| Original Image | 0.5465 | 1.9624 | 3.9509 | 4.0586 | 8.1117 | -5.0875 | 8.4133 |
| Inverted Image | 0.5465 | 1.9624 | 3.9509 | 4.0586 | 8.1117 | -5.0875 | 8.4133 |
| Rotated Image | 0.5464 | 1.9620 | 3.9506 | 4.0580 | 8.1096 | -5.0860 | 8.4166 |
| Half-size Image | 0.5466 | 1.9693 | 3.9464 | 4.0548 | 8.1037 | -5.0860 | 8.4053 |
| Max Value | 0.5466 | 1.9693 | 3.9509 | 4.0586 | 8.1117 | -5.0860 | 8.4166 |
| Min Value | 0.5464 | 1.9620 | 3.9464 | 4.0548 | 8.1037 | -5.0875 | 8.4053 |
| Abs Difference | 0.0002 | 0.0073 | 0.0045 | 0.0038 | 0.0080 | 0.0015 | 0.0113 |

ments as follows:

$$
\begin{aligned}
M_1 &= (\eta_{20} + \eta_{02}), \\
M_2 &= (\eta_{20} - \eta_{02})^2 + 4\eta_{11}^2, \\
M_3 &= (\eta_{30} - 3\eta_{12})^2 + (3\eta_{21} - \eta_{03})^2, \\
M_4 &= (\eta_{30} + \eta_{12})^2 + (\eta_{21} + \eta_{03})^2, \\
M_5 &= (\eta_{30} - 3\eta_{12})(\eta_{30} + \eta_{12})[(\eta_{30} + \eta_{12})^2 - 3(\eta_{12} + \eta_{03})^2] \\
&\quad + (3\eta_{21} - \eta_{03})(\eta_{21} + \eta_{03})[3(\eta_{30} + \eta_{12})^2 - (\eta_{21} + \eta_{03})^2], \\
M_6 &= (\eta_{20} - \eta_{02})[(\eta_{30} + \eta_{12})^2 - (\eta_{21} + \eta_{03})^2] \\
&\quad + 4\eta_{11}(\eta_{30} + \eta_{12})(\eta_{21} + \eta_{03}), \\
M_7 &= (3\eta_{21} - \eta_{03})(\eta_{30} + \eta_{12})[(\eta_{30} + \eta_{12})^2 - 3(\eta_{21} + \eta_{03})^2] \\
&\quad - (\eta_{30} - 3\eta_{12})(\eta_{21} + \eta_{03})[3(\eta_{30} + \eta_{12})^2 - (\eta_{21} + \eta_{03})^2].
\end{aligned}
\tag{5.6}
$$

#### 5.1.2.1   Hu's Invariant Moments

Figure 5.2 shows the original image, inverted image, image rotated by 9 degrees and a half sized image. The seven Hu moments of each of the four images are computed using Equation 5.6, and listed in Table 5.1. A log transform is taken to get meaningful values. The absolute difference between maximum value of Hu moments of all four images and minimum value of all four images is computed. It is oberved that the absolute difference is much less for all seven moments. Hence, Hu moments are invariant to object scaling, position, and orientation.

### 5.1.3   Human Activity Recgnition

In this subsection, we discuss how MHI and Hu moments can be used for recognizing five basic human actions like hand waving, sitting, getting up, boxing, and jogging. But it is view dependent; that is, the camera need to be at a certain angle. We classify each of the MHIs using a three-level classifier process:

1. We first calculate the set of seven Hu moments for each MHI and use it for classification. We will call it Cla1.

2. We calculate the horizontal and vertical projections of the MHI and then calculate the bias of MHI along the different directions (vertical and horizontal) with respect to the centroid of MEI. We use this for the second-level classification. We will call it Cla2.

3. We calculate the horizontal and vertical displacements between the centroids of the MHI and MEI. We will call it Cla3.

4. We use the following steps for the classification:

a. If both Cla1 and Cla2 agree, then the common result indicating the identified action is considered.

b. If there is a disagreement between Cla1 and Cla2, then we find the results using Cla3. The action that is supported by at least two of the three feature vectors is called the identified action.

c. If all three features indicate separate sets of actions, then we consider the action indicated by Cla1 as the valid result because Hu moments are the best indicators, and give better identification results than Cla2 and Cla3.

**FIGURE 5.2**: Invariance property of Hu moments: a) Original image, b) Inverted image, c) Image rotated by 9 degrees, d) Half-size image

### 5.1.3.1    Classification Using Hu Moments

Now we are able to obtain the Hu moments of the MHIs. We now move towards the recognition and classification of the MHIs into the five categories.

1. Before calculating the seven Hu moments, we normalize each of the MHIs.
2. We calculated the Hu moments for each of the templates and obtained the Hu moment matrix for the three sets of representative MHIs.
3. We formulated a distance measuring technique to calculate the distance between MHI of the input video and the three representative MHIs.
4. The similarity measure used is as follows:

$$I(A, B) = \sum_{i=1}^{7} |m_i^A - m_i^B| \tag{5.7}$$

where,

$$m_i^A = sign(h_i^A).log_{10}h_i^A \tag{5.8}$$

and

$$m_i^B = sign(h_i^B).log_{10}h_i^B \tag{5.9}$$

5. Before differencing, we take the absolute values of the seven moments. Also, this scaling was required because the values of the moments, especially the later ones, are very small. Hence, in implementing this, we have first considered three moments instead of all seven, since maximum information is contained in the initial moments.
6. This function $I(A, B)$ was calculated between the representative MHIs of the three actions and the MHI of the input video sequence.
7. The minimum of the three differences was taken and the corresponding action was established. The video was then annotated accordingly to display the result of the algorithm.

### 5.1.3.2    Projection and Displacement Features

The second-level classifier Cla2 is obtained as follows: If the R is number of rows and C is number of columns in the MHI, then we can obtain the horizontal and vertical profiles as follows: The vertical profile is

$$P_v[i] = \sum_{j=1}^{C} MHI[i, j] \qquad i = 1 \; to \; R \tag{5.10}$$

$$P_h[i] = \sum_{j=1}^{R} MHI[i, j] \qquad i = 1 \; to \; C \tag{5.11}$$

Using the above two vectors, we calculate the projection profile-based features as follows:

$$Cla2 = [\frac{\sum_{i=1}^{hct} P_h[i]}{\sum_{i=hct+1}^{C} P_h[i]} \quad \frac{\sum_{i=1}^{vct} P_v[i]}{\sum_{i=vct+1}^{R} P_v[i]}] \tag{5.12}$$

This indicates the bias of the obtained MHI with respect to the centroid of MEI. In a way, this feature relates to the temporal information of motion along both the directions. The third feature that we use as our classification is the shift between MEI and MHI centroids. It is obtained as follows:

$$Cla3 = [MHI_{xc} - MEI_{xc} \quad MHI_{yc} - MEI_{yc}] \tag{5.13}$$

The centroids of MHI and MEI differ because in MHI we have time information, too. So in a way, the above feature vector indicates the direction of motion of the centroid for any action sequence. Now, using the hierarchy of feature vectors that we discussed earlier, we perform the action recognition and classification.

### 5.1.3.3 Experimental Discussion

Figure 5.3 shows the successful recognition results obtained by using MHI and Hu moments for recognizing five basic human actions such as hand waving, sitting, getting up, boxing, and jogging. For these experiments, we have used a fixed CCTV camera. From a different viewpoint, we may not get the desired results. It gives good results for small variation in size but we need to use Hu

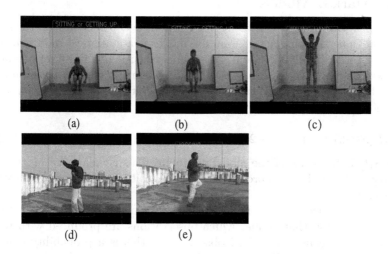

(a)   (b)   (c)

(d)   (e)

**FIGURE 5.3**: Human activity recognition results: (a) Sitting, (b) Getting up, (c) Hand waving, (d) Boxing, (e) Jogging

moments for better results. This method is illumination independent, which means it gives good results even under different lighting conditions. For this method, the background should be static, which means the camera should be at a fixed position. This method cannot handle the occlusion problem. The MEI template created during occlusion will be much different from the normal template. So, we cannot recognize the action that is being performed.

---

## 5.2    Hidden Markov Models

The hidden Markov model (HMM) is a statistical tool used for modeling time series data. It has been applied extensively for solving a large number of problems in the area of speech recognition, natural language processing, biomedical signal processing [9], and sport video processing [35], [27]. Recently, it has been adapted to human activity recognition as well, since the HMM tool is suitable for recognizing time sequential feature information. Andrei Markov gave his name to the mathematical theory of Markov processes in the early twentieth century [1]. But the main theory of HMM was developed by Baum and his colleagues in the 1960s [12]. In this section we will discuss basic concepts of HMMs, and three fundamental problems in HMM and use of HMM for activity recognition.

### 5.2.1    Markov Models

This is a process in which the state at any time depends only on the state at the previous time instant. A Markov model is a statistical model for describing time series of obervations. The model consists of a finite set of states. Thus, we can model state sequences using these state transition probabilities. The probability of a transition from state $i$ to state $j$ depends only on state $i$, and not on the earlier history, as follows:

$$P(S(t)|S(t-1), S(t-2), S(t-3), .....) = P(S(t)|S(t-1)) \qquad (5.14)$$

where $S(t)$ is the state of the model at time $t$. This is called the Markovian property of the model. The probability of all transitions out of any given state must sum to 1.

The states collectively model a set of observations. Each state has a probability distribution that defines which observations are produced with what probability. For continous valued observations, this is a probability density function. This is also called the emission or observation probability for the state. A full probabilistic description of the system requires specification of the current state, as well as all the previous states. A stochastic process could be called an observable Markov model because the output of the process is the set of states at each instant of time, where each state corresponds to an observable event.

## 5.2.2  Hidden Markov Models

An HMM is a sequence of finite states connected by state transitions. It begins in a designated initial state. In each discrete time step, a transition is taken into a new state, and then one output symbol is generated. The choice of transition and output symbol are both random, governed by probability distributions. The HMM can be thought of as a black box, where the sequence of output symbols generated over time is observable, but the sequence of states visited over time is hidden. Hence, this model is called hidden Markov model. Every state of HMM can be described by two types of probabilities: state transition probability (A) and symbol observation probability (B). In Figure 5.4, $S_1$, $S_2$ are the states of HMM and $V_1$ and $V_2$ are observable symbols. The state tranisition probability from state $i$ to state $j$ is given as

$$a_{ij} = P(S_i(t-1) \rightarrow S_j(t)) \qquad (5.15)$$

and

$$\sum_j a_{ij} = 1; \qquad \forall i \qquad (5.16)$$

The visible symbol probablity of state $j$ for symbol $k$ is given as

$$b_{jk} = P(V_k|S_j) \qquad (5.17)$$

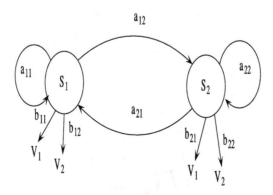

**FIGURE 5.4**: HMM state transition probabilities

and

$$\sum_k b_{ij} = 1; \qquad \forall j \tag{5.18}$$

### 5.2.2.1  Final State

HMM has a specific hidden state called a final state, the receiving state, or accepting state. Once HMM reaches the final state $S_0$, it cannot come out of that state. The state transition diagram including final state is shown in Figure 5.5.

## 5.2.3  The Three Fundamental Problems of HMM

There are three basic problems that must be solved in order to apply this HMM model for any real-world application. These problems are as follows:
1. Likelihood Evaluation
Given an HMM $\theta$ specified by $S$, $V$, $a_{ij}$, $b_{jk}$ and an observation sequence $V^T$, we have to find $P(V^T/\theta)$. This problem is called the likelihood evaluation problem, and the algorithm used is called the forward algorithm.
2. State Sequence Decoding
Given an HMM model $\theta$, and an observation sequence $V^T$, we have to find the most likely HMM state sequence for $\theta$ and $V^T$.
3. HMM Parameter Estimation
Given an HMM rough structure (i.e., set of states and transition structure)

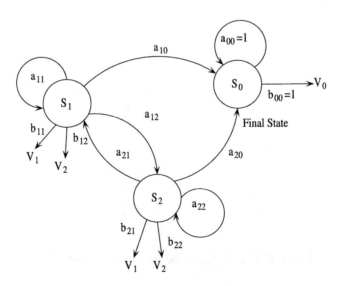

**FIGURE 5.5**: HMM state transition probabilities with final state $S_0$

and some labeled training data, estimate the HMM parameters that maximize the likelihood of the training data. This is the forward-backward algorithm, useful for training an HMM.

---

**Algorithm 9** Forward Algorithm

---

**Input:** $a_{ij}, b_{jk}, V^T, \alpha_j(0)$
**Output:** $P(V^T|\theta)$

1: initialize: $t \leftarrow 0, a_{ij}, b_{jk}, V^T$
2: Calculate Forward Probability
3: $\alpha_j(t) \leftarrow 0$, t=0 $j \neq$ initial state
4: $\alpha_j(t) \leftarrow 1$, t=0 $j =$ initial state
5: **for** each time $t$ **do**
6: $\quad t \leftarrow t+1$
7: $\quad \alpha_j(t) \leftarrow \sum_i [\alpha_i(t-1) \times a_{ij}] \times b_{jkv(t)}$
8: **end for**// $t = T$
9: Return $P(V^T|\theta) \leftarrow \alpha_0(T)$ for final state

---

### 5.2.3.1 Likelihood Evaluation

Given the HMM model $\theta$ and symbol sequence $V^T$, we have to estimate $P(V^T|\theta)$.

$$P(V^T|\theta) = \sum_{r=1}^{rmax} P(V^T|S_r^T)P(S_r^T) \tag{5.19}$$

where, $r$ indicates one of the possible state sequences and
$S_r^T = \{S(1), S(2), ....S(T)\}$.
Let the total number of hidden states be $N$, then $rmax = N^T$.

$$P(V^T|S_r^T) = \prod_{t=1}^{T} P(V(t)|S(t)) \tag{5.20}$$

$$P(S_r^T) = \prod_{t=1}^{T} P(S(t)|S(t-1)) \tag{5.21}$$

Substituting values from Equations 5.20 and 5.21 into Equation 5.19, we get

$$P(V^T|\theta) = \sum_{r=1}^{rmax} \prod_{t=1}^{T} P(V(t)|S(t))P(S(t)|S(t-1)) \tag{5.22}$$

The computation complexity for implemenation of Equation 5.22 is on the order of $N^T T$. In order to reduce computation, the forward Algorithm 9 is used.

### 5.2.3.2 State Sequence Decoding

In this problem, $V^T$ is given and we have to find the most probable sequence of hidden states using Algorithm 10.

---

**Algorithm 10** State Sequence Decoding Algorithm

---

**Input:** $a_{ij}, b_{jk}, V^T, \alpha_j(0)$
**Output:** Most probable state sequence $S^T$

1: initialize: $path \leftarrow \{ \}, t \leftarrow 0, j \leftarrow 0$
2: **for** each time $t$ **do**
3:     $t \leftarrow t+1$
4:     **for** each time $j$ **do**
5:         $j \leftarrow j+1$
6:         $\alpha_j(t) \leftarrow \sum_{i=1}^{N} [\alpha_i(t-1) \times a_{ij}] \times b_{jkv(t)}$
7:     **end for**// $j = N$
8:     $\hat{j} \leftarrow argmax_j \alpha_j(t)$
9:     Append $\hat{S}_j$ to path
10: **end for**// $t = T$
11: Return  path

---

**Algorithm 11** Baum Welch Algorithm

---

**Input:** Set of states S, Set of symbols V
**Output:** $a_{ij}, bjk$

1: initialize $\leftarrow a_{ij}, b_{jk}, p_i$, v
2: Calculate  Forward  Probability
3: $\alpha_j(t) \leftarrow 0$,  t=0  j $\neq$ initial state
4: $\alpha_j(t) \leftarrow 1$,  t=0  j = initial state
5: $\alpha_j(t) \leftarrow \sum_j [\alpha_i(t-1) \times a_{ij}] \times b_{jkv(t)}$  otherwise
6: Calculate  Backward Probability
7: $\beta_i(t) \leftarrow 0$, $w_i(t) \neq w_0$, t $\neq$ T
8: $\beta_i(t) \leftarrow 1$, $w_i(t) = w_0$, t = T
9: $\beta_i(t) \leftarrow \sum_j [\beta_i(t+1) \times a_{ij}] \times b_{jkv(t+1)}$  otherwise
10: Calculate $v_{ij}(t) \leftarrow \frac{[\alpha_i(t-1) \times a_{ij} \times b_{jkv(t)}] \times \beta_j(t)}{P(v^T|\theta)}$
11: Update state transition matrix and emission matrix
12: $a_{ij} = \leftarrow \frac{\sum_{t=1}^{T} v_{ij}(t)}{\sum_{t=1}^{T} \sum_k v_{ik}(t)}$
13: $b_{jk} \leftarrow \frac{\sum_{t=1}^{T} \sum_l v_{jl}(t)_{v(t)=v_k}}{\sum_{t=1}^{T} \sum_l v_{jl}(t)}$
14: Repeat till convergence

---

### 5.2.3.3   HMM Parameter Estimation

In order to train HMM, we must optimize $a$ and $b$ with respect to the HMM's likelihood of generating all of the output sequences in the training set, because this will maximize the HMM's chances of also correctly recognizing new data. Unfortunately, this is a difficult problem and it has no closed form solution. The best that can be done is to start with some initial values for $a$ and $b$, and then iteratively, the values of $a$ and $b$ are modified by reestimating them, until some stopping criterion is reached. This general method is called estimation-maximization (EM). A popular instance of this general method is the forward-backword algorithm, which is also called Baum Welch algorithm. The detailed steps are mentioned in the Algorithm 11 .

## 5.2.4   Limitations of Hidden Markov Models

1) There is no theoretically correct way for choosing the number of states. In an HMM classifier, a higher number of states does not necessarily imply better performance.
2) The next stage is only dependent on the current stage. So, this model is poor at capturing long-range correlation between the observed variables.

---

## 5.3   HMM-Based Activity Recognition

Human activity can be defined as a set of poses varying in time. In a real-world scenario, we can classify human activity as normal and abnormal based on a situation when activity is performed and the type of activities. So, we have to define a set of activities that can be called normal activity in that particular context. In a highly crowded scenario like an airport, we have a number of security cameras. It is not possible to monitor all the cameras at one time. Also, the maximum amount of time a person can concentrate on a video at a stretch is 20 minutes. It is most likely that a person will miss some abnormal activity due to human stress error. There is a need for an automatic activity recognition system that can reduce human error and increase security levels in various important places such as airports, railway stations, and bus-stops.

HMM can be used to model dynamic systems. It is used for classifying human activity as the activity can be described as a group of poses varying with time. Different features used for classifying human activity are shape-based feature, optical flow-based feature and appearance-based feature. Shape-based features are more sensitive to noise because the edges of the regions of interest are sensitive to noise. This feature set is not rotation invariant. Optical flow features are the velocity-based features, which are rotation invariant. In order to get optimal classification, both features are combined and used for training purposes. Figure 5.6 shows the block diagram of an HMM-based human activity system. With this proposed approach [32], we have developed HMM models for the following five set of activities: side gallop, two-hand wave, walk, run, bend. In this approach, the multi-frame averaging method is used for background extraction, and combined optical and shape-based features are used for human activity recognition.

## 5.3.1    Shape-Based Features

Initially, video frames were extracted and converted to gray-level images. Multi-frame averaging was used for background extraction. Background was subtracted from video frames, resulting in foreground extraction. Based on the centroid point, the ROI was segmented from video frames and edges were

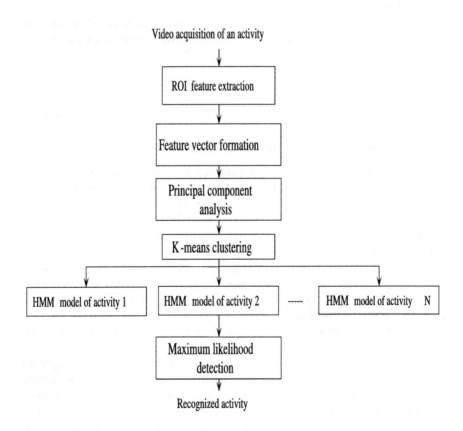

**FIGURE 5.6**: HMM-based human activity system

extracted using the Canny edge detector. Distance was calculated from the centroid to the edges and discrete Fourier transform (DFT) coefficients were extracted from it. An initial 20 DFT points were preserved because the most energy gets concentrated in the initial coefficients. Principal component analysis (PCA) was used for reducing the dimension of feature vector.

### 5.3.1.1 Discrete Fourier Transform

Discrete Fourier transform converts time domain information to frequency domain information. The resultant coefficients are complex numbers called coefficients of DFT. This function is usually implemented in software and hardware using the fast Fourier transform (FFT) algorithm. Mathematically DFT can be represented as:

$$X(K) = \sum_{n=0}^{N-1} x(n) \times \exp^{\frac{-2\pi k n}{N}} \tag{5.23}$$

The initial 20 DFT coefficients were preserved and others were deleted.

### 5.3.1.2 Principal Component Analysis

This is a method to identify similarities and differences in a dataset. Once the similarity is found, data can be compressed to lower dimensional data. Steps for PCA:
1. Collect information from various sources.
2. Subtract mean from all points to make data zero mean data.
3. Calculate the co-variance matrix.
4. Calculate eigen value and eigen vector of co-variance matrix
5. Order eigen vectors according to decreasing order of their eigen values.
6. The eigen vectors of highest eigen values are preserved and the lower components are rejected.
Because some components can be retained and some can be neglected, this results in lowering the dimension. In this case, PCA was applied on 20 DFT coefficients, the initial 8 PCA components were preserved and others were rejected. This results in a lower-dimension feature set of matrix order $8 \times 8$.

### 5.3.1.3 K means Clustering

K means clustering is a popular clustering algorithm that partitions the data into a number of sections having the nearest mean point. Here K indicates the number of clusters to be created. The algorithm steps can be described as follows:
1. Choose K initial cluster.
2. Compute point-to-cluster distances of all observations to each centroid.
3. Assign each observation to the cluster with the closest centroid.
4. Recompute the centroid for each cluster.
5. Repeat 2 to 4 until the centroids do not change.

In this case, k means was used to generate symbol sequence for the given feature set. This symbol sequence is used to train the HMM model.

### 5.3.2   Optical Flow-Based Features

Shape-based features are ROI dependent and are not rotation invariant. As the orientation of a person changes, the shape will change and so will the features. Hence, to make system rotation invariant and more efficient, optical flow based features are included in the feature set in addition to the shape-based features. The Lucas Kanade optical flow method was used to calculate velocity of an object.

### 5.3.2.1   Lucas Kanade Optical Flow Method

The Lucas Kanade method is the most popular method used for optical flow estimation. Here, the optical flow is assumed constant in the local neighborhood of the pixel under consideration. By combining information from several nearby pixels, the Lucas Kanade method can often resolve the inherent ambiguity of the optical flow equation. This method is less sensitive to image noise than point-wise methods. The local optical flow velocity vector $v_x, v_y$ must satisfy the following equation:

$$I_x(q_1) \times v_x + I_y(q_1) \times v_y = -I_t(q_1) \tag{5.24}$$

The above equation can be written in the form

$$A \times V = b \tag{5.25}$$

where:

$$A = \begin{bmatrix} I_x(q_1) & I_y(q_1) \\ I_x(q_2) & I_y(q_2) \\ ..... & ..... \\ I_x(q_n) & I_y(q_n) \end{bmatrix}$$

$$b = \begin{matrix} -I_t(q_1) \\ -I_t(q_2) \\ .... \\ -I_t(q_n) \end{matrix}$$

$$V = \begin{matrix} v_x \\ v_y \end{matrix}$$

where $I_x$ and $I_y$ are partial derivatives of an image in the x and y directions. The solution is obtained by:

$$V = (A^T \times A)^{-1} \times A^T \times b$$

$$\begin{pmatrix} v_x \\ v_y \end{pmatrix} = \begin{pmatrix} \sum_i I_x(q_i)^2 & \sum_i I_x(q_i) \times I_y(q_i) \\ \sum_i I_y(q_i) \times I_x(q_i) & \sum_i I_y(q_i)^2 \end{pmatrix} * \begin{pmatrix} -\sum_i I_x(q_i) \times I_t(q_i) \\ -\sum_i I_y(q_i) \times I_t(q_i) \end{pmatrix}$$

The matrix $A^T \times A$ is often called the structure sensor of the image.

### 5.3.2.2 Optical Features

The following optical flow-based features are extracted:
1. Velocity in x and y directions
2. Direction of flow
3. Vorticity
4. Divergence
5. Gradient tensor features
6. Total x and y velocity

Divergence
This is a scalar quantity. Mathematically, it can be represented as:

$$f(x,t) = \frac{\partial u(x,t_i)}{\partial x} + \frac{\partial v(x,t_i)}{\partial y} \tag{5.26}$$

It is a summation of the partial derivative of velocity in the x and y directions. The physical significance of the divergence is that it captures the amount of expansion taking place in the fluid. Here the motion concept is global and represents motion of an independent body part.

Vorticity
This is a measure of local spin around the axis perpendicular to the plane of flow field. It represents rigidity in the flow and is useful in representing local motion in a body. Mathematically it can be represented as:

$$f(x,t) = \frac{\partial v(x,t_i)}{\partial x} - \frac{\partial u(x,t_i)}{\partial y} \tag{5.27}$$

Gradient Tensor Features
Small-scale structures present in the flow field are called eddies, and can be characterized by a large velocity gradient.

$$\partial u(x,t_i) =$$

$$\begin{pmatrix} \frac{\partial u(x,t_i)}{\partial x} & \frac{\partial u(x,t_i)}{\partial y} \\ \frac{\partial v(x,t_i)}{\partial x} & \frac{\partial v(x,t_i)}{\partial y} \end{pmatrix}$$

$$R(x,t_i) = -det(\partial u(x,t_i))$$

**TABLE 5.2**: State Transition Matrix

**TABLE 5.3**: Bending

| | | | |
|---|---|---|---|
| 0.2155 | 0.7845 | 0 | 0 |
| 0 | 0.2154 | 0.7846 | 0 |
| 0 | 0 | 0.9373 | 0.0627 |
| 0 | 0 | 0 | 1 |

**TABLE 5.4**: Boxing

| | | | |
|---|---|---|---|
| 0.5 | 0.5 | 0 | 0 |
| 0 | 0.6586 | 0.3414 | 0 |
| 0 | 0 | 0.9782 | 0.0218 |
| 0 | 0 | 0 | 1 |

**TABLE 5.5**: Clapping

| | | | |
|---|---|---|---|
| 0.3484 | 0.6516 | 0 | 0 |
| 0 | 0.3484 | 0.6516 | 0 |
| 0 | 0 | 0.9787 | 0.0213 |
| 0 | 0 | 0 | 1 |

**TABLE 5.6**: Handwave

| | | | |
|---|---|---|---|
| 0.5753 | 0.4247 | 0 | 0 |
| 0 | 0.5753 | 0.4247 | 0 |
| 0 | 0 | 0.9792 | 0.0208 |
| 0 | 0 | 0 | 1 |

**TABLE 5.7**: Walking

| | | | |
|---|---|---|---|
| 0.8571 | 0.1429 | 0 | 0 |
| 0 | 0 | 1 | 0 |
| 0 | 0 | 0.9056 | 0.0944 |
| 0 | 0 | 0 | 1 |

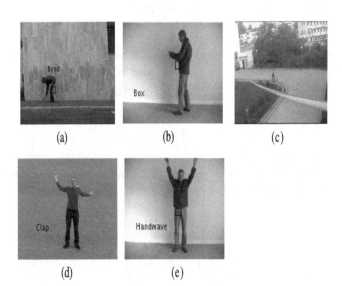

**FIGURE 5.7**: HMM-based activity recognition: (a) Bending, (b) Boxing, (c) Walking, (d) Clapping, (e) Hand-waving

### 5.3.3  Implementation and Results

HMM is used for modeling time sequential activity. The Baum Welch algorithm is used for training purposes. Once the classifier is trained, the test videos are tested by extracting features from the video and classifying the video based on maximum likelihood estimation. Here, we have designed an HMM model for five different activities: bending, boxing, clapping, hand-wave, and walking. The trained state transition matrixes for bending, boxing, clapping, hand-wave, and walking activities are given in Tables 5.3, 5.4, 5.5, 5.6, and 5.7 respectively. The overall accuracy achieved was 90%. The applied techniques found to be scale and direction invariant due to fusion of both optical and shape-based features. The optical feature is direction invariant and the shape feature extracted is normalized to make it scale invariant. Snapshots of the successfully recognized clips are shown in Figure 5.7.

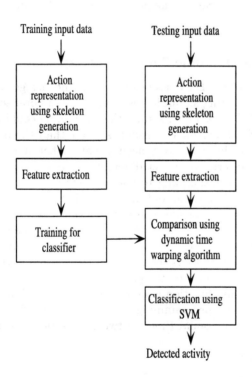

**FIGURE 5.8**: Block diagram of DTW-based activity recognition

## 5.4    Dynamic Time Warping-Based Activity Recognition

For human activity recognition, the dynamic time warping (DTW) approach is very useful because of its robustness against variation in speed or style in performing action. DTW would be applied to the feature vector created from the skeleton joints motion. Further, the rapid development of inexpensive depth sensors, e.g., Microsoft Kinect, provides better accuracy for real-time body tracking for activity recognition applications. Hence, for improving the rate of recognition, a dataset would be created using Kinect to recover human joints, body-part information in a 3-D real-world coordinate system. Classification of certain human actions into walking, kicking, punching, and handshaking classes would be performed using the well-known SVM classifier. The block diagram of the proposed approach is shown in Figure 5.8.

### 5.4.1    What Is Dynamic Time Warping?

In time series analysis, DTW is one of the well-known algorithms for detecting patterns by measuring similarity between two temporal sequences, which may vary in speed. For example, similarities in running could be detected using DTW, even if one person was running faster than the other. DTW has gained its popularity by being very accurate and efficient in measuring similarity between time-series to detect different shapes and patterns. In DTW, two feature vectors corresponding to two sequences are aligned by *warping* the time axis iteratively until an optimal match between the two sequences is found. The best alignment between these two sequences is given by a warping path that minimizes the overall distance between them. The overall distance involves finding all possible routes, and for each route, computing the overall distance. The overall distance is the minimum of the sum of the distances between the individual elements on the path divided by the sum of the weighting function. The weighting function is used to normalize path length.

DTW uses dynamic programming as a tool and yields optimal warping path in $O(m \times n)$ polynomial time for two time series of lengths m and n, respectively.

$A = (a_1, a_2, , a_i, .., a_m)$

$B = (b_1, b_2, , b_j, .., b_n)$

First of all, a local distance matrix is calculated using these two time series.

$$Distance - matrix(i, j) = |A_i - B_j| \qquad (5.28)$$

Then, a local cost matrix $C$ is created by evaluating the cost matrix for all the elements of $X$ and $Y$. From this cost matrix, the minimum cost warping path (lowest distance) is to be obtained.

A warping path is a sequence $W = (w_1, ..., w_{|w|})$ where for $k \in \{1, ..., |w|\}$,

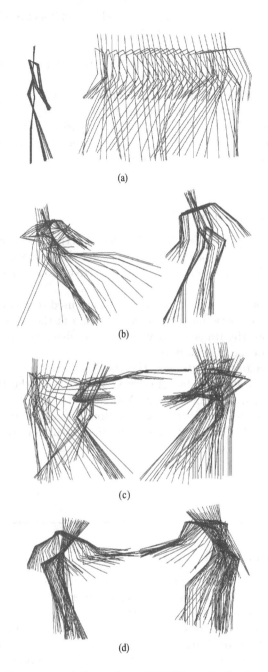

**FIGURE 5.9**: Framewise skeleton for (a) Walking, (b) Kicking, (c) Punching, and (d) Handshaking

$w_k = c(a_k, b_k)$, satisfying the following conditions:

1) Boundary condition: The starting and ending symbols are aligned to each other for both time series. $w_1 = c(a_1, b_1)$

$w_{|w|} = c(a_m, b_m)$

2) Monotonicity condition: The symbols are aligned in monotonously non-decreasing order.

3) Continuity condition: No observation symbols are to be skipped.

This leads to an overall cost function $C(W) = \sum w_k$, $k = 1$ to $k = |w|$. The function $C(W)$ denotes all possible warping paths between two sequences $X$ and $Y$. The DTW algorithm is used to find the warping path that gives the lowest distance measure between two series.

### 5.4.2    Implementation

The following steps are used for DTW-based activity recognition [53]:

1) Each frame from the depth camera is considered as a feature vector.

2) For a given video sequence, frames are converted into a sequence of feature vectors.

3) Joint orientation of each body part's joints are used to construct a feature vector. The distance between the respective joints in the consecutive frames is computed for all the frames in a video. These distances are considered to be a feature vector for that activity.

4) For action recognition purposes, the feature vectors of the video under testing is compared with the list of defined actions using the DTW method. Classification is then performed using SVM.

Figure 5.9 shows the framewise skeleton generated using the joint-coordinates of the bodies of all frames in the videos for walking, kicking, punching, and handshaking activities.

---

## 5.5    Abnormal Activity Recognition

Humans perform various types of activities. Thses activities are classified as normal or abnormal based on the contexts. The video data from the various cameras contain different numbers of persons performing various activities. The number of persons in a given scene is an important parameter for the human activity recognition system. Next, vehicles can also perform different activities on the road on behalf of humans. The abnormal activities are recognized in order to control traffic. As the number of persons in the video increases, it is very difficult to recognize the activities of individual persons.

In that case, the activity of the whole group is recognized and examples of such activities are dispersion and gathering.

In a human activity recognition problem, the interactions can be classified as:
1) Single-person or no interaction
2) Few-persons interaction
3) Heavy-crowd interaction
4) Person-vehicle interaction
5) Person-facility interaction
6) Heavy crowd-facility interaction.

1) Single person or no interaction

This refers to scenarios where a single person may or may not be interacting with his environment. Occlusion hardly occurs, as there is no interaction with any other person, and one may go for simple trajectory analysis for behavior prediction. This problem is the simplest one. Examples of single person activities are standing, loitering, jogging, talking on a mobile phone, trespassing on a property, stealing, and playing.

2) Few-persons interaction

This refers to scenarios when a couple or group of people interact with each other or with their environment. Since few persons are involved in the interaction, occlusions may occur. Hence, trajectory analysis for behavior prediction becomes difficult. The examples are shaking hands, exchanging things, playing, fighting, attacking. Figure 5.9 (b), (c), and (d) show kicking, punching, and hand-shaking activities that belongs to this category.

3) Heavy crowd interaction

This refers to scenarios where there is a heavy crowd or traffic. Trajectory analysis for behavior prediction is almost impossible for these scenarios because of severe occlusion. In such a situation, a head count is done.

4) Person and vehicle interaction:

This type of interaction refers to scenarios such as driving a car, coming out, getting into a car, drunken driving, vehicle vandalism.

5) Person and facility interaction

Under this type of interaction, people interacting with a location or facility can be modeled. The examples of this type of interaction are behavior of a person in a bank, elevator, ATM, and checkpoints. Any abnormal behavior can be deemed as suspicious in this case.

6) Heavy crowd and facility interaction

These kinds of situations are very complex, and examples of such scenarios are passengers coming out or getting into a train, moving along the platform, or movement in the subway.

## 5.6    Challenges of Intelligent Human Activity Recognition

To date, several techniques have been proposed for human activity recognition. But many of these techniques work succcessfully on specific datasets, and fail to produce similar results in more complex, large-scale datasets. Some of the challenges of surveillance systems are:

1) Poor video quality due to changes in illumination, moving cameras, high speed of action execution, complex background.

2) How to handle occlusion, and shadows.

2) Highly discriminative features are required for recognizing human activity.

3) Features should be independent of viewpoint; otherwise the performance of surveillance systems will suffer as the viewpoint changes.

4) A criminal can fool the algorithm of surveillance systems by posing as performing normal activity.

5) Many sources of noise, such as the shaking of leaves, must be extracted from videos with unconstrained backgrounds.

6) Very large areas, such as border areas, playgrounds, and road squares, cannot be monitored by a single camera. A network of cameras that coordinate with one another and share information is required. Cameras at different orientations with overlapped fields of view give video data that may have different illumination, or backgrounds. So, multi-view information needs to be combined properly.

## 5.7    Summary

In this chapter, various methods of human activity recognition are discussed. The concept of MHI and activity recognition based on MHI is discussed. The importance of Hu moments is presented by showing implementation details. The HMM-based activity recognition approach is discussed in depth, providing implementation details. There are other classifiers, such as Bayesian Belief Networks [40], and SVM-based classifiers, which can be used for activity recognition. After detecting human activity, this activity can be classified as normal or abnormal activity. The DTW-based approach for activity recognition is also discussed, because of its robustness against variation in speed or style in performing actions. DTW would be applied to the feature vector created from the skeleton-joint's motion.

# Chapter 6

---

## Video Object Tracking

---

## 6.1 Introduction

### 6.1.1 What Is Video Object Tracking?

Video object tracking is the process of estimating the positions of a moving objects over time using a camera. Tracking creates the trajectory of an object in the image plane as it moves in the scene. A consistent label can be assigned by the tracker to the moving objects in different frames of a video. A tracker can also provide additional information about moving objects such as orientation, area, and shape. However, it is a challenging task because of the loss of information caused by projection of the 3-D world onto a 2-D image, occlusion, noise, changes in illumination, and complex object motion. Also, most of the applications have real-time processing requirements which is also a challenge. Video tracking is used in human activity recognition, human-computer interaction, video surveillance, video communication, video compression, traffic control, and medical imaging. Since there is increased demand for automated video analysis, a lot of research is going on in the area of object tracking [22].

### 6.1.2 Tracking Challenges

Since video contains large amounts of data, video object tracking is a time-consuming process. Further, for tracking objects, there is a need to recognize the object from the video frame. Object recognition is also a challenging problem because the object may change shape, size, location, and orientation over subsequent video frames. Video tracking can be especially difficult when the objects are moving with high speed relative to the frame rate. For these situations, usually a motion model is developed, and this will describe how the object will move to different positions in the successive frames. Various algorithms and schemes have been introduced in the last few decades, and these are discussed in this chapter.

The performance of the algorithm changes because of several challenges that are mentioned in Figure 6.1. One of the main challenges is when the background appears similar to the object of interest or other objects that are present in the scene. This phenomena is called clutter. Another type of challenge is variation of appearance because of sensor noise, object pose, illumination changes in the scene, or occlusion. Since the object is moving, the appearance of the object may vary its projection on a video frame plane. Also, in the acquisition process of video, it may be possible to introduce a certain amount of noise during the acquistion of video signal. Sometimes moving objects may get occluded behind other objects that are present in the scene. In such cases, that video object tracker may not observe the object of interest.

---

**Algorithm 12** Video Object Tracking Algorithm

---

**Input:**  Input video sequence
**Output:**  Output video sequence with trajectories

  1:  Choose a feature in initial frame.
  2:  Choose a feature space.
  3:  Represent model in the selected feature space.
  4:  Select an ROI around target location in current frame.
  5:  Find the most similar candidate based on similarity function and mark its centroid.
  6:  Repeat for all frames in the video.

---

### 6.1.3    Steps of Video Object Tracking System

In the development of any algorithm, we need to have suitable object representation, the right features, and a good tracking algorithm. The general steps of tracking are given in Algorithm 12. Video tracking involves identifying the object and labeling it properly. The main steps of the tracking systems are as follows:

1. Background identification

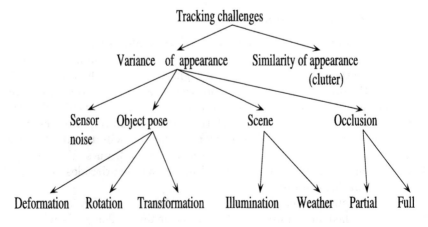

**FIGURE 6.1**: Main challenges in video object tracking

2. Foreground object detection
3. Object labelling
4. Handling occlusion problem

### 6.1.3.1 Background Identification

For identification of background, there are various methods that can be employed. One of the methods is based on Gaussian mixture models (GMM). A GMM is a parametric probability density function. It is represented as a weighted sum of Gaussian component densities. GMM can be used to model the colors of an object, which will help to perform real-time color-based object tracking. By using the GMM-based background model, frame pixels are deleted from the video to achieve the desired tracking. It has a disadvantage that if there is much less motion by the foreground object, it will start treating it as a background object. As shown in Figure 6.2, it can be noticed that when humans become stationary for quite some time, they eventually become a part of the ever-updating background and the track is discontinued. When they start moving again, the track continues.

### 6.1.3.2 Foreground Object Detection

The foreground objects can be extracted by subtracting the background frame from the current frame. First, the background frame is converted from color frame to gray-scale image. Then the gray-scale background image is subtracted from the gray-scale current frame. This image is then converted to

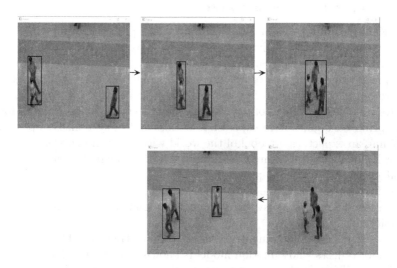

**FIGURE 6.2**: GMM-based background modeling

a binary image using a proper threshold. The morphological operators such as erode and dilation are performed on the binary image to remove noise. In the output image, white color represents foreground objects and black color represents background.

### 6.1.3.3   Object Labeling

After identifying the foreground objects, labels should be assigned to these objects. For this, the contours around the object are identified. Based on some criteria such as contour of extracted objects, the labels are assigned during the subsequent frames.

### 6.1.3.4   Handling the Occlusion Problem

When a foreground object goes behind another object, we cannot see it. This problem is known as an occlusion problem. In video object tracking, occlusions significantly undermine the performance of any tracking algorithms. The Kalman filter deals with this problem by predicting the path of the occluded object using dynamic features such as displacement, velocity, and acceleration. If many objects possess similar features, then the Kalman filter is unable to differentiate between the different objects.

---

## 6.2   Kalman Filter

### 6.2.1   What Is a Kalman Filter?

A Kalman filter is an optimal recursive data-processing algorithm that uses measured data to estimate the state of a system. It was primarily developed by the Hungarian engineer Rudolf Kalman. As in Bayesian theory, when predicting future events, we not only include our current experiences, but also our past knowledge. Sometimes, that past knowledge is so good that we have a very clear model of how things should pan out. The Kalman filter is an optimized quantitative expression of this kind of system. By optimally combining an expectation model of the world with prior and current information, a Kalman filter accurately estimates how things will change over time. The algorithm is a two-step process:
1) The first step predicts the state of the system,
2) The second step uses noisy measurements to refine the estimate of the system state.
Kalman filters are predominantly used for object tracking. They are more suitable for objects whose motion model is known. Along with that, they also assimilate addtional information to reveal the next object position more robustly. The main objectives of using a Kalman filter are as follows:

(1) Predicting future location for objects.
(2) Reducing noise due to inaccurate detections.
(3) Associating multiple objects to their tracks.

### 6.2.2 How Does a Kalman Filter Work?

Consider an imperfect detector, such that it is prone to false positives. This implies that the detector might not be able to detect objects every time. The imperfection further implies that it is unable to provide us with the exact position and scale. Moreover, the execution of this detector is costly.

Suppose we want to track a single moving object. To be able to robustly track it, we would have to make the best use of the information about the object once it is detected. The detector gives us the location of the object. Object Motion Model:
To predict the next position of the object, we need an object motion model, such as constant velocity motion and constant acceleration motion.
Measurement noise:
The imperfection of the detector implies that there will be noise in object locations, generally referred to as measurement noise.
Process noise:
The chosen motion model will also have noise in predicting object motion. This is referred to as process noise.
To predict the next object position, we consider three parameters: Object motion model, measurement noise, and the process noise. So, the object will be redetected efficiently, and we can cope with object occlusion.

### 6.2.3 Kalman Filter Cycle

A Kalman filter basically propagates and updates Gaussians and their covariances. It first predicts the next state with the help of the state transition (e.g., motion) model. This is followed by noise measurement information in the correction phase. The cycle is then repeated.
Step 1: Initial State
Here, we state the initial state values of position and velocity, along with the Gaussian covariance matrix depicting the initial uncertainty in order to start the tracking process.
Step 2: Predict
In this step, the next state is predicted along with updating the uncertainty of object state, i.e., state prediction and covariance prediction. We use the state transition matrix and process noise covariance for predicting next state.
Step 3: Correct
In Kalman filter update, the correction is done based on the noisy measurement information. Kalman gain (K) specifies how much we pay attention to

our prediction versus the actual measurement.

### 6.2.4   Basic Theory of Kalman Filter

The Kalman filter algorithm is one of the point tracking algorithms. It uses state space equations and measurement equations for tracking. So, object tracking with a Kalman filter uses prediction and correction equations. First, the object location is predicted, and then in the next frame, using measurement formulas, the predicted location will be corrected, and a better estimation of the object location will be obtained.

#### 6.2.4.1   Prediction Equations

$$x_k = F_k x_{k-1} + B_k \vec{u_k} + w_k$$
$$P_k = F_k P_{k-1} F_k^T + Q_k \tag{6.1}$$

Here, $x_k$ is the present state, $x_{k-1}$ is the state at time k-1, and $P_k$ is the state transition matrix that applies the effect of the each of the system state variables (e.g., velocity, position). $B_k$ is the control input matrix that applies the effect of each control input parameter (e.g., acceleration, force), $u_k$ is the vector containing any control inputs (e.g., acceleration, force), $P_k$ is the covariance matrix of the state vector, and $Q_k$ is the covariance matrix of external noise. $w_k$ is an additive white Gaussian noise.

To start Kalman filter prediction equations, we have to select the system model (i.e., constant velocity model, constant acceleration model) of the dynamic system and the associated state variables. Prediction equations for the constant velocity and constant acceleration models are described below.
a) Constant velocity model
The prediction equation for the constant velocity model is given by the following equation:

$$x_k = \begin{pmatrix} p_k \\ v_k \end{pmatrix} = \begin{bmatrix} 1 & \Delta t \\ 0 & 1 \end{bmatrix} = F_k x_{k-1} \tag{6.2}$$

So, $u_k = 0$, as there are no external influences like acceleration or force.

$$P_k = cov(x_k) = F_k P_{k-1} F_k^T \tag{6.3}$$

These prediction equations are based on basic laws of kinematics, as given below:

$$p_k = p_{k-1} + \Delta t v_{k-1} + 1/2a\Delta t^2$$
$$v_k = v_{k-1} + a\Delta t \tag{6.4}$$

where $p_k$ and $v_k$ are the position and velocity after time.

b) Constant acceleration model
The prediction equations for the constant acceleration model are given by

$$x_k = F_k x_{k-1} + B_k \overrightarrow{u_k} + w_k$$
$$P_k = F_k P_{k-1} F_k^T + Q_k \tag{6.5}$$

where $F_k = \begin{bmatrix} 1 & \Delta t \\ 0 & 1 \end{bmatrix}$, $B_k = \begin{pmatrix} \Delta t^2/2 \\ \Delta t \end{pmatrix}$ and $Q_k$ is the covariance matrix of external noise, $w_k$. These equations are based on the basic kinematics equations, which are given below:

$$p_k = p_{k-1} + \Delta t v_{k-1} + 1/2 a \Delta t^2$$
$$v_k = v_{k-1} + a \Delta t \tag{6.6}$$

c) Prediction equation for 2-D objects
The prediction equations for 2-D objects are similar to those for a 1-D object. Here, the state equations contain both the horizontal and vertical components of state variables.

$$X = \begin{bmatrix} x \\ y \\ x' \\ y' \end{bmatrix}, F_k = \begin{bmatrix} 1 & 0 & \Delta t & 0 \\ 0 & 1 & 0 & \Delta t \\ 0 & 0 & 1 & 0 \\ 0 & 0 & 0 & 1 \end{bmatrix}, B_k = \begin{bmatrix} \Delta t^2/2 & 0 \\ 0 & \Delta t^2/2 \\ \Delta t & 0 \\ 0 & \Delta t \end{bmatrix} \tag{6.7}$$

The calculation of the above-mentioned matrices are based on the following kinematics equations:

$$x_k = x_{k-1} + \Delta t x'_{k-1} + 1/2 x''_{k-1} \Delta t^2$$
$$y_k = y_{k-1} + \Delta t y'_{k-1} + 1/2 y''_{k-1} \Delta t^2$$
$$x'_k = x'_{k-1} + x'_{k-1} \Delta t$$
$$y'_k = y'_{k-1} + y'_{k-1} \Delta t \tag{6.8}$$

### 6.2.4.2 Update Equations

The Kalman filter model updates the state variables of the model by the following equations:

$$x'_k = x_k + K'(z_k - H_k x_k)$$
$$P'_k = P_k - K' H_k P_k$$
$$K' = P_k H_k^T (R_k + H_k P_k H_k^T) \tag{6.9}$$

**Algorithm 13** Basic Algorithm of Kalman Filter

**Input:** Input image
**Output:** Output image with trajectories

1: Define the motion model, i.e., constant velocity model, constant acceleration model.
2: Define the state variables and state matrix of the motion model.
3: Find $F_k$ and $B_k$ from the kinematics equation. Set the value of the $Q_k$ matrix.
4: Predict the state matrix $x_k$ and its covariance $P_k$.
5: Find the sensor matrix $H_k$ and set a small value for covariance of measurement noise $R_k$.
6: From steps 4 and 5, find the Kalman gain ($K'$).
7: Update the state matrix $x_k$ and its covariance $P_k$ depending on the Kalman update equation
8: Repeat steps 2 to 7 with present updated state matrix as previous state matrix.

Here, $x'_k$ is the updated state matrix, $K'$ is the Kalman gain, which depends on state covariance matrix, $P_k$, sensor matrix, $H_k$, and measurement noise $R_k$. Kalman gain is the metric that determines whether measurement or prediction is more reliable, and it moves the updated values towards the less noisy part. The basic flow chart of a Kalman filter is shown in Figure 6.3. The detailed steps are available in Algorithm 13.

### 6.2.4.3 Measurement Equations

$$z_k = Cx_k + n_k \qquad (6.10)$$

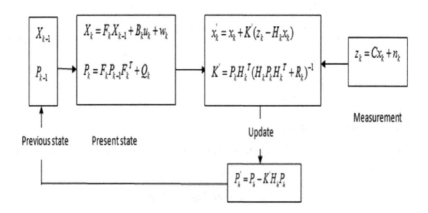

**FIGURE 6.3**: Block diagram of Kalman filter

Here C is the measurement matrix. It fuses all the state variables with a predefined relationship. The measurement matrix C is similar to the sensor matrix, H.

$$(\mu_0, \Sigma_0) = (H_k x_k, H_k P_k H_k^T) \tag{6.11}$$

where $\mu_0$ and $\Sigma_0$ are the expected mean and covariance of the prediction data. H is the sensor matrix, which is similar to measurement matrix C.

## 6.2.5  Implementation

The Kalman filter is an optimal recursive filter that uses a series of incomplete and noisy measurements to estimate the state of a dynamic system. Some of the applications include a control, navigation, computer vision, and time-series econometrics. This example illustrates how the Kalman filter is used for tracking objects, and incorporates the following:

a. Predicting future location of object

b. Reducing noise due to inaccurate detections

c. Associating multiple objects to their tracks

During multiple object tracking, a new Kalman filter is instantiated for each new object. The filter is used in order to predict the position of the object in the next frame. At a point of time, the extracted list of ROIs is compared with the predicted list of objects from the previous frame. If a merged object is detected, the system is not able to retrieve a new observation vector and the Kalman filter is updated only by using the previous state vector. This approach can be followed only if the motion of the objects in the scene is uni-

**FIGURE 6.4**: Tracking results before applying Kalman filter

form, i.e., the speed is constant. If the acceleration of the considered object is not zero, the prediction error of the Kalman filter increases and it causes a substantial tracking failure. Figures 6.4 and 6.5 show tracking of persons in the video data recorded by us on the IIT Patna campus. As shown in Figure 6.4 (c) and (d), the person was not tracked because of an occlusion problem. In order to solve this problem, we used Kalman filter whose results are shown in Figure 6.5.

## 6.3    Region-Based Tracking

In region-based tracking, the deviations of the image sections are used for tracking moving objects. For instance, a cross-correlation function is used to detect vehicle blobs in a vehicle detection problem. The motion sections can be identified using background subtraction. M. Kilger [25] proposed this method for a real-time traffic monitoring system for moving vehicle detection. Congested traffic further poses the problem of occluded vehicles. In many research methodologies, a Gaussian distribution of pixel values is used to model both a human body and a background scene. However, it is very difficult to model a body part of the person such as head and hand. C. R. Wren et al. [62] proposed a real-time system called pfinder (person finder) that solves the problem for tracking a single person using a fixed camera. To remove shadows,

**FIGURE 6.5**: Tracking results after applying Kalman filter

a combination of color and gradient information obtained from background subtraction is used. Color clues are used to distinguish between the objects during occlusion. Some researchers have proposed a color-based tracking system instead of human body tracking [46]. Since the visual parts depend on the color of a person's clothing, the tracker will fail if two persons in the group have similar dress.

## 6.4 Contour-Based Tracking

The algorithm for contour-based tracking is shown in Figure 6.6. It uses the boundary contour of a moving and deforming object to track it. Initially, the contour of an object is extracted. In the consequent frames, the system automatically outlines the contours of the objects at video rate. It is useful in many areas, such as automated surveillance, vehicle tracking, and motion-based recognition.

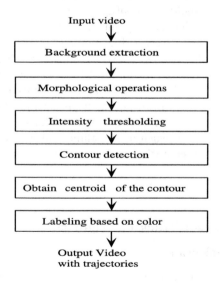

**FIGURE 6.6**: Flowchart for contour-based tracking algorithm

## 6.5    Feature-Based Tracking

This performs tracking by matching features between images, which are obtained by extracting elements from objects and clustering them into higher-level features. It is an unsupervised tracking algorithm. In general, feature-based tracking methods can adapt successfully and rapidly, to allow real-time tracking of multiple objects. Feature-based tracking algorithms are broadly classified as follows:

1) Global feature-based methods
2) Local feature-based methods
3) Dependence-graph-based methods

Global feature-based methods use features such as color, centroid, and perimeters, whereas local feature-based algorithms use features such as line segments, curve segments, and corner vertices. After we have extracted the feature points, for a feature point in a given frame, the algorithm finds a corresponding point in the next frame. To find the corresponding point, the sum-of-squared-difference (SSD) criterion is used where the displacement is calculated by minimizing the SSD. The next step is the validation of whether the corresponding point gives a large value under the Harris criterion. If it does not satisfy the threshold, we identify a candidate set of feature points in a small neighborhood of the target point and repeat the above process. The algorithm may output no corresponding point due to disappearance of the target point. It is capable of tracking feature points and detecting occlusions.

A feature-based tracking algorithm must first assume a form to model the motion of objects. The translational motion model, which dominantly concentrates on regions undergoing linear movement, is used most of the time. With increase in complexity of motion and geometric deformations, it is necessary to use other complex models, such as the affine motion model and combined motion model. Object segmentation is also used to isolate features of interest for tracking.

### 6.5.1    Feature-Based Tracking Algorithm

#### 6.5.1.1    Feature Selection

Good features are necessary for a feature-based algorithm to work. In this algorithm, candidate features in one or more frames are selected. For selection of point features, the most common algorithm is the Harris corner detector. In this algorithm, we first choose a point $x$, and calculate the quality of the point with

$$C(x) = det(G) + k \times trace^2(G) \tag{6.12}$$

computed on the window $W(x)$, and $k$ is the constant value that can be selected by the user. In the equation, $G$ is a $2 \times 2$ matrix that is given by

$$\begin{bmatrix} \sum I_x^2 & \sum I_x I_y \\ \sum I_x I_y & \sum I_y^2 \end{bmatrix}$$

where $I_x$ and $I_y$ are the gradients obtained by convolving the image $I$ with the derivatives of a pair of Gaussian filters. To ensure that a feature is selected only if the window contains sufficient texture [41], a point $x$ is selected as a feature if $C(x)$ exceeds a certain threshold $\tau$. In general, to achieve high tracking efficiency, we do not expect the features to be concentrated in a small region within the image. If a specific region is rich in terms of feature points, multiple feature points may be selected, which may result in inefficient tracking. To solve this problem, we define a minimal space $p$ between two selected features. This minimal space will ensure whether the candidate feature point is sufficiently distanced from other selected feature points.

### 6.5.1.2 Sum-of-Squared-Difference Criterion

Since we have assumed a simple translational deformation model, to track a point feature $x$, we search for the new location $x + \delta x$ on the frame at time $t + \tau$ whose window is most similar to the window $W(x)$. To measure similarity, the sum-of-squared-difference (SSD) criterion can be used. It compares the image window $W$ centered at the location $(x, y)$ at time $t$ and another new location $(x + dx, y + dy)$ on the frame at time $t + dt$, where the point could have moved between the two frames. The displacement $d$ is obtained by minimizing the SSD criterion

$$E_t(dx, dy) = \sum [I(x + dx, y + dy, t + dt) - I(x, y, t)]^2 \qquad (6.13)$$

where the subscript $t$ indicates the translational deformation model. An alternative way to compute the displacement $d$ is to evaluate the function at each location and choose one that gives the minimum error. This formulation was proposed by Lucas and Kanade in the context of stereo algorithms [49]. Later, Tomasi and Kanade [58] modified this algorithm in a more general feature-tracking problem.

### 6.5.1.3 Pyramidal Decomposition

In image processing, multi-scale decomposition is used to improve efficiency. Generally, the recursive form of the algorithm decomposes signals into information at different levels. Simoncelli and Freeman [55] proposed a steerable pyramid for efficient and accurate linear decomposition of an image into scale and orientation. J. Bouguet [63] developed an algorithm to implement pyramidal image scales of the Lucas-Kanade feature tracker. In the tracking algorithm proposed in [18], a pyramid of images is constructed through smoothing and down sampling of the original image. For instance, let $I_0$ be the original image and $I_{L-1}$ represent the image at level $L1$. The $L^{th}$ level image is then defined as follows:

$$
\begin{aligned}
I^L(x,y) = \frac{1}{4}I^{L-1}(2x, 2y) &+ \frac{1}{8}(I^{L-1}(2x-1, 2y) + I^{L-1}(2x+1, 2y) \\
&+ I^{L-1}(2x, 2y-1) + I^{L-1}(2x, 2y+1)) \\
&+ \frac{1}{16}(I^{L-1}(2x-1, 2y-1) + I^{L-1}(2x+1, 2y+1) \\
&+ I^{L-1}(2x-1, 2y-1) + I^{L-1}(2x+1, 2y+1)) \quad (6.14)
\end{aligned}
$$

For a given feature central point $x$, its corresponding coordinates on the pyramidal images are computed as

$$
x^L = \frac{x}{2^L} \quad (6.15)
$$

Then we compute each motion direction vector $d^L$ on each level of pyramidal images. Finally, we could sum up all levels of motion vector $d$ as

$$
d = \sum 2^L d^L \quad (6.16)
$$

A small tracking window will be selected to preserve the details in the image and improve accuracy, while a big window size will be selected for handling larger motions and to improve robustness. Hence, an important issue with a feature-based tracker is the tradeoff between accuracy and robustness. The advantage of pyramidal implementation is that while each motion vector $d^L$ is obtained by way of smaller integration windows, the overall displacement vector $d$ can account for larger pixel motions, thus achieving both accuracy and robustness. Algorithm 14 explains the pyramidal implementation that ensures both accuracy and robustness [18].

---

## 6.6    Model-Based Tracking

To obtain the location and pose of the observer, model-based tracking is used if we know the model of the environment. It has applications in augumented reality and robot navigation systems. Model-based tracking uses a 3-D object in the scene for estimating the pose of a camera. It relies on natural objects in the scene. Model-based tracking can be classified into two categories, namely recursive tracking and tracking by detection.

---

**Algorithm 14** Feature-Based Tracking Algorithm

---

**Input:** Input image
**Output:** Output image with trajectories

  1: Function Selection (Image I)
  2: Choose window $W_I(x)$
  3: Calculate G as follows
  4: $G = \begin{bmatrix} \sum I_x^2 & \sum I_x I_y \\ \sum I_x I_y & \sum I_y^2 \end{bmatrix}$
  5: Calculate threshold $\tau$
  6: Sort $x$ in the decreasing order of $C(x)$
  7: **if** $C(x_i) > \tau$ **then**
  8:    **if** $x_i - x_j > \rho$ **then**
  9:       Include $x_i$ as feature point
10:    **end if**
11: **end if**
12: Function Tracking (x)
13: Build a pyramid of $k$ levels of images
14: **for** each level $k$ **do**
15:    $d^k = -G^{-1}b$ *for image pair* $(I_1^k, I_2^k)$
16:    Move the window $W(x)$ by $2d^k$ through warping the next level image $I^{(k-1)}2(x)$
17:    $d = d + 2d^k$
18:    $k = k - 1$
19: **end for**
20: Calculate $C(x)$ as follows
21: $C(x) = det(G) + k \times trace^2(G)$
22: **if** $C(x_i) > \tau$ **then**
23:    $x_i = x_{i+1}$
24: **else**
25:    Find all candidate features in the tracking window $W_2(x)$ on image $I_2$.
26:    Apply SSD criteria.
27:    **if** $min(SSD) < \tau_e$ **then**
28:       $x_i = x_{i+1}$
29:    **else**
30:       return empty
31:    **end if**
32: **end if**

---

In recursive tracking, the previous camera pose is used as an estimate to calcuate the current camera pose, wheras tracking by detection computes current pose without any prior knowledge, thus being computationally expensive and demanding high processing power.

---

**Algorithm 15** Simple KLT Tracking Algorithm

**Input:** Input image
**Output:** Output image with trajectories

1: Detect Harris corner in the first frame.
2: For each Harris corner, compute motion between consecutive frames.
3: Link motion vectors in successive frames to get a track for each Harris point.
4: Introduce new Harris points by applying Harris detector after some frames (say 15 frames).
5: Track new and old Harris points using steps 1 to 3

---

## 6.7    KLT Tracker

The Kanade-Lucas-Tomasi tracker, popularly known as the KLT tracker, is a feature tracker. This tracker is based on the early work of Lucas and Kanade [48], which was fully developed by Tomasi and Kanade [58], and explained clearly in the paper by Shi and Tomasi [54]. KLT uses spatial intensity

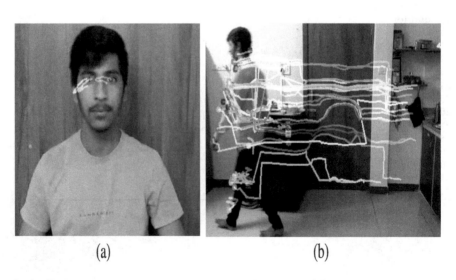

(a)                                        (b)

**FIGURE 6.7**: Tracking results of the KLT tracker. (a) Result on small motion of object. (b) Result on walking video sequence. (See ebook for color.)

information for searching the position in the next frame that gives the best match. It is faster than traditional techniques for finding the potential matches between the images. KLT is based on the Harris corner detector, which is a mathematical operator that finds features in an image. It is simple to compute, and invariant to rotation, scale and illumination changes. The simple KLT tracker is explained in Algorithm 15. Figure 6.7 (a) shows KLT results for small motion such as eye and head movement, and Figure 6.7 (b) shows the result of a KLT tracker on a walking video sequence. The goal of this algorithm is to match a template image $T(x)$ to an input image $I(x)$. The $I(x)$ could also be a small window within an image.

The set of allowable warps is $W(x; p)$, where $p$ is a vector of parameters. For translations

$$W(x; p) = \begin{bmatrix} x + p_1 \\ y + p_2 \end{bmatrix} \tag{6.17}$$

The best alignment minimizes image dissimilarity.

$$\sum_x [I(W(x; p)) - T(x)]^2 \tag{6.18}$$

is a non-linear optimization! The warp $W(x; p)$ may be linear but the pixel values are in general non-linear. Hence, in the modified problem, it is assumed that some $p$ is known and best increment $\Delta p$ is sought.

$$\sum_x [I(W(x; p + \Delta p)) - T(x)]^2 \tag{6.19}$$

is solved with respect to $\Delta p$. When found, then $p$ gets updated

$$p \leftarrow p + \Delta p \tag{6.20}$$

$$\sum_x [I(W(x; p + \Delta p)) - T(x)]^2 \tag{6.21}$$

is linearized by using first-order Taylor expansion.

$$\sum_x [I(W(x; p + \Delta p)) + \nabla I \frac{\partial W}{\partial p} \Delta p - T(x)]^2 \tag{6.22}$$

where

$$\nabla I = [\frac{\partial I}{\partial x}, \frac{\partial I}{\partial y}] \tag{6.23}$$

is the gradient image computed at $W(x; p)$. The term $\frac{\partial W}{\partial p}$ is the Jacobian of the warp.

Differentiate

$$\sum_x [I(W(x; p + \Delta p) + \nabla I \frac{\partial W}{\partial p} \Delta p - T(x)]^2 \qquad (6.24)$$

with respect to $\Delta p$

$$2 \sum_x [\nabla I \frac{\partial W}{\partial p}]^T [I(W(x; p + \Delta p) + \nabla I \frac{\partial W}{\partial p} \Delta p - T(x)] \qquad (6.25)$$

and setting Equation 6.25 equal to zero yields

$$\Delta p = H^{-1} \sum_x [\nabla I \frac{\partial W}{\partial p}]^T [T(x) - I(W(x; p))] \qquad (6.26)$$

where $H$ is the Hessian matrix

$$H = \sum_x [\nabla I \frac{\partial W}{\partial p}]^T [\nabla I \frac{\partial W}{\partial p}] \qquad (6.27)$$

### 6.7.1    Limitations of the KLT Tracker

1) The KLT tracker does not guarantee that the corresponding point in the next frame is a feature point, because KLT tracker only uses the Harris criterion for the first frame but not for the other frames.
2) KLT may not handle occlusion well.

---

**Algorithm 16** Tracking Algorithm with Mean Shift

---

**Input:** Target Model $q_u$ and its location $y_0$ in initial frame
**Output:** Location $y_1$ in the next frame

1: Initialize the location at the current frame with $y_0$
2: Compute $\{p_u(y_0)\}$ and $\{\rho[p_u(y_0), q]$
3: Compute the next location $y_1$ using mean shift
4: Compute $\{p_u(y_1)\}$ and $\{\rho[p_u(y_0), q]\}$
5: While $\{\rho[p_u(y_1), q] < \{\rho[p_u(y_0), q]\}$
6: Do $y_1 = \frac{1}{2}(y_0 + y_1)$
7: If $\|y_1 - y_0\|$ is small enough, stop
8: Else set $y_0 = y_1$ and repeat from step-2

---

## 6.8 Mean-Shift-Based Tracking

### 6.8.1 What Is Mean Shift?

Mean shift is a non-parametric feature-space analysis technique that uses the given discrete data sample of a density function to locate its maxima [6], [5]. It is an iterative mode detection algorithm in the density distribution space. It uses kernels to compute the weighted average of the observations within the tracking window and repeats this computation until convergence is attained at a local density mode. Thus, density modes can be located without explicitly estimating the density.

### 6.8.2 Algorithm

Initially the color feature of the tracking window is divided in $u$ colors. The target object mode can be computed as follows:

$$q_u = C \sum_{i=1}^{n} k(\|\frac{x_i - x_0}{h}\|^2) \delta[b(x_i) - u] \qquad (6.28)$$

where, $\{x_i\} = 1.......n$ are the pixel locations of the target window centered at $x_0$. $k$ is the kernel profile, which is convex and monotonic decreasing. $h$ is the kernel bandwidth. $b(x_i)$ associates to the histogram value corresponding to the color $x_i$. $\delta$ is Kronechker delta function. $C$ is a normalization constant that satisfies $\sum q_u = 1$. Similarly, the target candidate at location $y$ can be presented by

$$p_u(y) = C_h \sum_{i=1}^{n} k(\|\frac{y - x_i}{h}\|^2) \delta[b(x_i) - u] \qquad (6.29)$$

where $C_h$ is also a normalization constant that satisfies $\sum p_u = 1$. The similarity of $p_u(y)$ and $q_u$ can be measured by Bhattacharya's coefficient.

$$\rho[p(y), q] = \sum_{u=1}^{M} \sqrt{p_u(y) q_u} \qquad (6.30)$$

Maximizing the Bhattacharya coefficient leads to the following mean-shift iterations. The new estimate of the target position $y_1$ is calculated to be a weighted sum of pixels contributing to the model.

$$y_1 = \frac{\sum_{i=1}^{n} x_i w_i g(\|\frac{y - x_i}{h}\|^2)}{\sum_{i=1}^{n} w_i g(\|\frac{y - x_i}{h}\|^2)} \qquad (6.31)$$

where

$$g(x) = -k'(x) \tag{6.32}$$

$$w_i = \sum_{u=1}^{M} \sqrt{\frac{q_u}{P_u(y)}} \delta[b(x_i) - u] \tag{6.33}$$

Figure 6.8 shows tracking results on a video sequence of walking. The detailed steps are available in Algorithm 16.

### 6.8.3   Advantages

The advantages of the mean shift algorithm are as follows:
1. Mean shift is an application-independent tool suitable for real data analysis.
2. No assumptions of any predefined shape on data clusters are made.
3. Arbitrary feature spaces are handled.
4. Mean-shift algorithm is fast and performs satisfactorily for various sequences.
5. It is useful for clustering, mode seeking, probability density function, and tracking.

### 6.8.4   Disadvantages

The limitations of the mean shift algorithm are as follows:
1. The selection of a window size is not trivial. Inappropriate window size can cause modes to be merged, or generate additional shallow modes. In most cases, adaptive window size is used.
2. For fast moving objects, it does not track accurately. It may even lose the target.
3. Sometimes it gets stuck in a local minimum.
4. If we use global color histogram features, localization drift may be observed.

**FIGURE 6.8**: Mean shift tracking result. (See ebook for color.)

**TABLE 6.1**: Applications of Video Object Tracking

| Application | Role of Object Tracking |
|---|---|
| Motion-based recognition | Unusual human movement recognition such as loitering |
| Automated surveillance | Monitoring a scene to detect abnormal activities |
| Video indexing | Automatic labeling of videos and retrieval of the videos in large databases |
| Human-computer interaction | Gesture recognition based on eye gaze tracking for data input to the computers |
| Traffic control | Real-time traffic monitoring to avoid traffic-jam, and accidents |
| Vehicle navigation | Suggest best path to vehicles for avoiding obstacles |

## 6.9 Applications of Tracking Algorithms

Trajectory analysis is very useful for applications such as human activity recognition, video surveillance, video indexing and retrieval, human computer interaction, traffic monitoring, and vehicle navigation, as mentioned in Table 6.1. The detailed implementation of abnormal activity recognition using trajectories is discussed below.

### 6.9.1 Trajectory-Based Unusual Human Movement Recognition

A trajectory is the path that a person moves along as a function of time. The trajectory in a scene is often used to analyze the human activity. We have used a trajectory to determine whether the person's movement is abnormal or not [51]. We can detect activities such as loitering around severely vulnerable places like bank vaults, ATMs, and military installations, based on the loitering trajectories shown in Figure 6.9 (b). If somebody's movement matches with abnormal trajectories, then the algorithm declares that this person is loitering and security should be informed. Treading track algorithms are used to detect closed-path trajectories and spiral trajectories [66].

### 6.9.1.1    Closed Path Detection

As shown in Figure 6.10 (a), we draw line-segments $L_1, L_2, \ldots, L_N$ between $x_0, y_0$ and $x_1, y_1$, $x_1, y_1$ and $x_2, y_2$, and so on. Then we find the intersection between any two line-segments. If the line-segments intersect, then we mark the person's trajectory as being closed-path. Based on Algorithm 17, a closed-path event is detected as shown in Figure 6.10 (b).

### 6.9.1.2    Spiral Path Detection

As shown in Figure 6.11 (a), to detect spiral, the center of all points of path is calculated. After that divide region of all points in 8 regions assuming $(x_c, y_c)$ as center of region. Then we find the angle $\alpha(n)$ between point $(x_n, y_n)$ and center point $(x_c, y_c)$ using the following formula :

$$\alpha(n) = tan^{-1}\left(\frac{y_n - y_c}{x_n - y_c}\right) \tag{6.34}$$

We find the angle of all points, and then count how many points are lying in each region. If the number of points lying in each region exceeds a thresholded

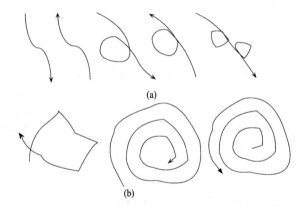

**FIGURE 6.9**: Trajectories of human movement: (a) Normal trajectories, (b) Loitering trajectories

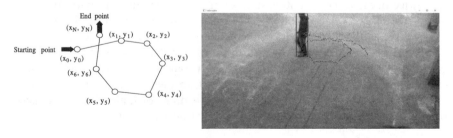

**FIGURE 6.10**: (a) Closed path trajectory, (b) Result

---

**Algorithm 17** Closed Path Detection

---

**Input:** Recorded video frames of spiral path activity
**Output:** Output video with trajectory and labeled as closed path

  1: Let the starting point of the path be $(x_0, y_0)$, and end point of the path be $(x_N, y_N)$ where $N + 1$ is the number of video frames corresponding to closed path activity.
  2: Draw line segment $L_1, L_2, ...., L_N$ by connecting the points
     $(x_0, y_0)$ *and* $(x_1, y_1)$, $(x_1, y_1)$ *and* $(x_2, y_2), ....(x_{N-1}, y_{N-1})$ *and* $(x_N, y_N)$,
  3: **for** each line-segment $n$ **do**
  4:    Check if line-segment $L_n$ intersects with $L_1, L_2, ....L_{n-1}$ line-segments
     If at any instance two line-segments cross each other,
     declare the path as closed path.
  5: **end for**// $N$ line-segments

---

**Algorithm 18** Spiral Path Detection

---

**Input:** Recorded video frames of spiral path activity
**Output:** Output video with trajectory and labeled as spiral

  1: Compute centroid of treading track of the object:

$$X_c = \frac{x_0 + x_1 + x_2 + ...... + x_N}{N + 1}$$
$$Y_c = \frac{y_0 + y_1 + y_2 + ...... + y_N}{N + 1}$$

     where $N + 1$ is the number of video frames corresponding to spiral path activity.
  2: Divide region of all points into eight equal regions assuming $(X_c, Y_c)$ as the center of the region
  3: **for** each video frame $n$ **do**
  4:    Find the angle between points and center point using the following formula:

$$\alpha(n) = tan^{-1}\left(\frac{y_n - y_c}{x_n - x_c}\right)$$

  5: **end for**// $N + 1$ frames
  6: **for** each video frame $n$ **do**
  7:    **for** each region $i$ **do**
  8:       Count how many points are lying in each region.
  9:    **end for**// 8 regions
 10:    If number of points lying in each region exceeds a threshold, label the trajectory as spiral.
 11: **end for**// $N + 1$ frames

---

average, then we label the trajectory as spiral. Based on Algorithm 18, a spiral path event is detected as shown in Figure 6.11 (b).

---

## 6.10    Summary

Video object tracking has wide applications in video surveillance, such as video compression, vision-based control, human computer interfaces, augmented reality, robotics, and traffic control. In this chapter, the general steps of a video tracking system, the basics of a Kalman filter, and its role in handling occlusion are discussed. Although Kalman filter is a renowned tool in object tracking and occlusion handling, its failure in complete occlusion is still an unsolved problem. Tracking approaches such as region-based tracking, contour-based tracking, feature-based tracking, mean shift-based tracking are discussed in brief. The success of a particular approach depends largely on the problem domain. Further, there exists a cost/performance trade-off. For real-time applications, we need a fast high-performance system. Bacause of the complexity of the tracking problem, there is room for improvement in various tracking algorithms. A typical application of trajectory-based abnormal activity recognition is presented in order to understand the implementation aspects of tracking algorithms.

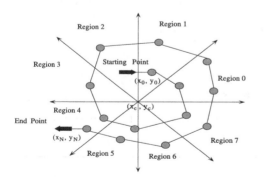

**FIGURE 6.11**: (a) Spiral path trajectory, (b) Result

# Part III

# Surveillance Systems

# Chapter 7

## Camera Network for Surveillance

Closed-circuit television (CCTV) cameras are widely used for video surveillance, to transmit a video signal to monitors in the control room. The different communication schemes used in a CCTV system are point to point (P2P), point to multipoint, and mesh wireless links. Siemens AG in Germany installed the first CCTV system in 1942. This CCTV system was designed and installed by German engineer Walter Bruch for observing the launch of V-2 rockets. Since CCTV cameras are now cheaply available due to rapid advances in technology, the use of CCTV cameras for surveillance systems has increased significantly. Nowadays, CCTV cameras are being installed for security purposes, be it in airports or railway stations. The use of cameras for such purposes is part of intriguing research that will bring improvements in the existing technology.

## 7.1 Types of CCTV Cameras

### 7.1.1 Bullet Camera

Bullet CCTV cameras are mainly designed for indoor applications. They can be mounted on a wall or ceiling. Since these cameras are placed inside bullet-shaped housing, they are called bullet cameras. The camera is designed to capture images from a fixed area. The pan tilt and zoom facility is not provided with this camera. The unit is mounted pointing at a fixed field of view, as shown in Figure 7.1 (a). They are used in residential as well as commercial locations.

### 7.1.2 Dome Camera

Since these cameras have dome-shaped housing, they are called dome cameras. Dome housings are designed to conceal the orientation of the camera. These units serve a dual purpose: People will know the facility is being watched and clients will feel that facility is being protected. Since camera orientation is hidden inside the casing, people often assume it is pointing in all directions.

For example, a fixed dome camera is covering a door at the end of a hallway, but people will assume that it will cover the entire hallway. In this way property will be protected. As shown in Figure 7.1 (b), some advaced units allow the camera to spin quickly within the housing. These types of cameras are often called speed domes. They are used in public places such as railway stations and bus stops.

### 7.1.3   Desktop Cameras

These are small-size cameras used for Skype and other low-resolution teleconference applications, as shown in Figure 7.1 (c). They can be eaily mounted to a desktop monitor. These cameras are also called board cameras. If we place cameras at discrete locations, they are called discrete cameras. The discrete housing is also designed to hide the entire camera. These can be mounted on a wall or ceiling to look like smoke detectors, motion sensors or nothing at all, with pinhole or flush mount lenses.

### 7.1.4   Vari-Focal Camera

A camera with a vari-focal lens allows the operator to zoom in or out while still maintaining focus on the image as shown in Figure 7.1 (d). This helps the operator to capture close-ups of activities at longer distances.

**FIGURE 7.1**: a) Bullet camera, b) Dome camera, c) Desktop camera, d) Varifocal camera, e) PTZ camera, f) IP camera

### 7.1.5  IP Camera

Internet protocol (IP) cameras transmit images over a computer network and the Internet, often compressing the bandwidth so as not to overwhelm the web. They are available in both hardwired and wireless versions. IP cameras are easier to install than analog cameras because they do not require a separate cable or power boost to send images over a longer distance. Videos can be seen of people using IP cameras anywhere and anytime regardless of the distance from the camera. Hence, IP cameras are preferred to provide long-distance surveillance soultions. A typical IP camera is shown in Figure 7.1 (f). The main components of an IP camera include a lens, an image sensor, one or several processors, memory, cable, switch, router, client, and server. The processors are used for image processing, compression, video analysis, and networking functionalities. A client is used to display and monitor the video. This network camera provides web server, FTP (file transfer protocol), and e-mail functionalities, and includes many other IP network and security protocols.

1) Front panel of a general network camera has IR LEDs, built-in microphone, built-in speaker, lens, wireless antenna.

2) Rear panel has LAN, DC power supply of 5V/2A, power light, network light, SD card, audio input jack, audio output jack, and RESET button. The LED will blink when power and network cable are plugged in. Audio input jack is used to plug an external microphone.

### 7.1.6  Wireless Camera

Not all wireless cameras are IP-based. Some wireless cameras can use alternative modes of wireless transmission. But no matter what the transmission method, the primary benefit to these units is still the same as with IP cameras. They are extremely easy to install.

### 7.1.7  PTZ Camera

Remotely positioned cameras are typically called PTZ cameras because of their ability to pan, tilt, and zoom in on a subject. The operator can move the PTZ camera left or right, up or down, and zoom the lens closer or farther away. The horizontal movement of the camera from left to right and right to left is called pan. The vertical movement of the camera from up to down or vice versa is called tilt.

These cameras are used in surveillance systems where there is an actual surveillance specialist monitoring the images. Also, pan-tilt-zoom functionality can be automoated on a timed basis as shown in Figure 7.1 (e). A PTZ camera may be set to automatically patrol an area, but they are oftentimes most effective when manually controlled. A higher zoom range for the PTZ camera is helpful when you are covering a large area with only one PTZ camera.

### 7.1.8    Infrared Cameras

These are also called night vision cameras. They are devices that form an image using infrared radiation. Infrared cameras (IR) have IR LEDs positioned around the outer edges of the camera lens, which gives the camera its night vision. Human vision capacity is limited to a very small portion of the electromagnetic spectrum. But thermal energy has a much longer wavelength than visible light, which the human eye can't even see. The portion of the spectrum we perceive is dramatically expanded by thermal imaging. This helps us to *see* thermal energy emitted from an object. In the infrared case, everything with a temperature above absolute zero emits heat. Even very cold objects such as ice cubes emit infrared radiation which helps us to see in highly lit and totally dark environments. A typical IR camera is shown in Figure 7.2.

The IR radiation emitted by an object depends on the object's temperature. If the object's temperature is high, greater IR radiations will be emitted. IR allows us to see what we cannot see with our eyes. IR cameras produce images of invisible infrared and provide precise non-contact temperature measurement capabilities. An IR camera detects infrared energy and converts it into an electronic signal. Then the electronic signal is processed to produce a thermal image on a video monitor and perform temperature calculations. Heat sensed by an IR camera can be very precisely measured, allowing us to monitor thermal performance. Infrared cameras are extremely cost-effective, and are used in many diverse applications. Some of the areas are as follows:
1) Security Systems
IR security cameras are practically useful and have fewer false alarms than other solutions. They can be used for boundary security purposes in defense to observe activities during the night.

**FIGURE 7.2**: IR camera

2) Maritime Systems

Professional mariners navigate safely around the clock with thermal night vision IR cameras. From small work boats to offshore commercial and passenger vessels, IR cameras take situational awareness to the next level, helping captains navigate safely, avoid obstacles and deter piracy.

3) Traffic Control Systems

IR cameras are used to control systems for traffic signals, roadway and tunnel incidents, and pedestrian crossings.

### 7.1.9   Day Night Cameras

Day night cameras are designed to compensate for varying light conditions, to allow the camera to capture images. In many cases, these camera units have a wide dynamic range, to function in glare, direct sunlight, reflections, and strong backlight at all hours. Hence, they are positioned in outdoor places such as parking lots, fence areas, and roadsides to monitor vehicles and human activities.

### 7.1.10   High-Definition Cameras

Ultra high-definition cameras are often used in niche markets such as casinos. These cameras give the operators the ability to zoom in with extreme clarity. These cameras have capability to transmit videos.

---

## 7.2   Smart Cameras

### 7.2.1   What Is a Smart Camera?

A smart camera is an intelligent camera that not only captures images but also is capable of extracting specific information from the captured images. It generates event descriptions or decisions that are used in intelligent video surveillance systems. A smart camera is a self-contained, stand-alone vision system with a built-in image sensor and all necessary communication interfaces.

## 7.2.2   Components of Smart Cameras

A smart camera usually consists of several of the following components: image sensor, image digitization circuitry, image memory, powerful processor such as a DSP processor, program and data memory (RAM, nonvolatile FLASH memory), communication interface (RS232, Ethernet), I/O lines (often opto-isolated), built-in lens or lens holder (usually C, CS, or M-mount), built-in illumination device such as LED, real-time operating system (e.g., VCRT), and optional video output (VGA or SVGA).

## 7.2.3   Why Do We Need a Smart Camera?

Smart cameras expand the vision capabilities of USB cameras. USB cameras can be converted into smart camera by incorporating an on-board processor to enhance vision capabilities while maintaining price, size, and power consumption.

## 7.2.4   Features

1) Powerful Smart Camera
These cameras handle precise data faster with better resolution. They also provide options for a dynamic field of vision. They have a high frame rate, and because of that, they have faster performance. They have a buffer on the camera. An on-camera preprocessor manages more data. Because of stand-alone mode, power consumption is low. There is no need for external computation.

2) Small Smart Camera
These cameras are also called USB smart cameras. The on-board processing capability of these cameras is consistent, and unaffected by computer operating system and software delays. Smart cameras can also be hooked into a single computer to create an integral system. There is no need for external computation. These small cameras have a modular design for minute-size dimensions, and they are suitable for hand-held or offsite devices.

3) Adaptable Smart Camera
These cameras have a USB connectivity option. They are simple to integrate and the cost is also low. Since USB technology is universally supported by almost any computer, these cameras do not require additional costly interface circuit boards. These cameras have various interface options, and they have battery options for stand-alone functionality.

## 7.3 Smart Imagers

### 7.3.1 What Is an Imager?

An imager is a sensor that detects and conveys the information that constitutes an image. It does so by converting the variable attenuation of waves (as they pass through or reflect off objects) into signals, the small bursts of current that convey the information. The waves can be light or other electromagnetic radiation. An imager is also called an image sensor.

### 7.3.2 Types of Imagers

There are two types of imagers: charged coupled device (CCD) imagers and complementary metal oxide semiconductor (CMOS) imagers.

#### 7.3.2.1 CCD Imager

A CCD imagers consists of a dense matrix of photodiodes. These photodiodes convert light energy in the form of photons into an electronic charge. Electrons generated by the interaction of photons with silicon atoms are stored in a potential well and can subsequently be transferred across the chip through registers and output to an amplifier. The basic operation of a CCD imager is as follows:

1) For capturing images, there is a photoactive region and a transmission region made out of a shift register.

2) An image is projected by a lens on the capacitor array (the photoactive region). Each capacitor of the array accumulates an electric charge proportional to the light intensity at that location.

3) A 1-D array used in cameras captures a single slice of the image, while a 2-D array used in video and still cameras captures a 2-D picture corresponding to the scene projected onto the focal plane of the sensor.

4) Once the array has been exposed to the image, a control circuit causes each capacitor to transfer its contents to its neighbor.

5) The last capacitor in the array dumps its charge into a charge amplifier, which converts the charge into a voltage.

6) By repeating this process, the controlling circuit converts the entire semiconductor contents of the array to a sequence of voltages, which it samples, digitizes into and stores in some form of memory.

Types of CCD Image Sensors

1. Interline transfer CCD image sensor

2. Frame transfer CCD image sensor

Nowadays, most CCD image sensors use interline transfer. The comparison of these two types of sensors is presented in Table 7.1.

**TABLE 7.1**: Comparison of Interline Transfer with Frame Transfer CCD Imager

| Interline Transfer CCD Image Sensor | Frame Transfer CCD Image Sensor |
|---|---|
| It uses optimized photodiodes with better spectral response. | It uses photogates. |
| It has lower fill factor. | It achieves higher fill factor with no photodiode. |
| The image is captured at the same time (snap shot operation). | Image is captured by mechanical shutter. |

#### 7.3.2.2   Complementary Metal Oxide Semiconductor

CMOS circuits use a combination of p-type and n-type metal oxide semiconductor field effect transistors (MOSFETs) to implement logic gates and other digital circuits. CMOS circuitry dissipates less power when static. It is denser than other implementations having the same functionality. In most CMOS devices, there are several transistors at each pixel that amplify and move the charge using wires. With the CMOS approach, each pixel can be read individually. In a CMOS sensor, each pixel has its own charge-to-voltage conversion, and the sensor often includes amplifiers, noise-correction, and digitization circuits so that the chip outputs digital bits. With each pixel doing its own conversion, uniformity is lower. The percentage of a pixel devoted to collecting light is called the fill factor of the pixel. CCDs have a 100% fill factor but CMOS cameras have much less. The lower the fill factor, the less sensitive the sensor is, and the longer the exposure times must be. Too low a fill factor makes indoor photography without a flash virtually impossible. A comparison of CCD and COMS imagers is presented in Table 7.2.

## 7.4   Multiple View Geometry

There are intricate geometric relations between multiple views of a 3-D scene. These relations have to do with the camera's motion, calibration, and scene structure. Several relationships exist between the images with different views. These are very important for the calibration and reconstruction from images. Many insights in these relationships have been obtained in recent years. Multiple view geometry is the subject where relations between coordinates of feature points in different views are studied. It is an important

**TABLE 7.2**: Comparison of CCD and CMOS Imagers

| CCD Imager | CMOS Imager |
| --- | --- |
| Many separate chips are required for other circuits. | It incorporate other circuits on the same chip. |
| The camera is bigger, heavier, and costly. | The camera is smaller, lighter, and cheaper. |
| It also requires more power. | It requires less power so batteries last longer. |
| It cannot switch between still photography and video. | It can switch modes on the fly between still photography and video. |
| CCD has a simple pixel and chip. | CMOS has more complex pixel and chip. |
| CCD performs better in low light conditions. | They suffer in low light conditions. |
| They suffer in high light conditions. | They excel in capturing outdoor pictures on sunny days. |
| CCDs have a 100% fill factor. | They have much lower fill factor. |
| Suitable for indoor photography. | Suitable for outdoor photography. |

tool for understanding the image formation process for several cameras using reconstruction algorithms.

## 7.5 Camera Network

### 7.5.1 What Is Camera Networking?

Camera networking is a concept that aims to study and analyze the communication process among various camera devices or networks of cameras that are linked together to exchange information and share resources. The cameras used in such camera networks are IP or USB cameras.

### 7.5.2 Camera Devices

There are various camera devices that can be used for the communication process, namely IP and USB cameras. An IP camera is a type of digital video camera that can send and receive data via a computer network and the Internet, as shown in Figure 7.3. Although most cameras that do this are webcams, the term *IP camera* is usually applied only to those used for surveillance. An IP camera is networked over a fast-ethernet connection and sends its signal to the main server via computer network or internet link. IP cameras are available in wired as well as wireless mode, and its maintenance cost is less than that of a traditional CCTV camera. The image quality acquired by IP cameras is better than that of traditional CCTV cameras. IP cameras support two-way communication. Hence, an alert signal can be sent to various places, in case of any suspicious activity. Hundreds of gigabytes of video and image data can be stored in video servers and can be retrieved at any time.

On the other hand, a USB webcam is a camera that connects to a computer, usually through plugging it into a USB port on the machine. The video is fed to the computer, and using an Internet video file can be transmitted. Because webcams rely on a computer to connect to the Internet, they must have a dedicated computer that is turned on at all times in order to function. Hence, IP cameras are preferred over USB cameras.

---

## 7.6     Camera Calibration

Camera calibration is the process of estimating parameters of the camera using images of a special calibration pattern. The parameters include camera intrinsics, distortion coefficients, and camera extrinsics. We can use these parameters to correct for lens distortion, measure the size of an object, or determine the location of the camera in the scene. Multi-camera calibration (MCC) maps different camera views to a single coordinate system. In many surveillance systems, MCC is a key pre-step for other multi-camera-based analysis. It is the process of estimating intrinsic and/or extrinsic parameters.

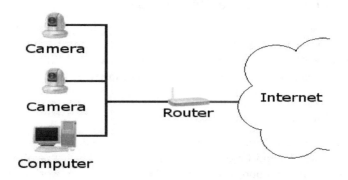

**FIGURE 7.3**: IP camera networking

Intrinsic parameters deal with the camera's internal characteristics, such as its focal length, skew, distortion, and image center. Extrinsic parameters describe its position and orientation in the world. Extrinsic parameters are used in applications such as to detect and measure objects, navigation systems for robots, and 3-D scene reconstruction.

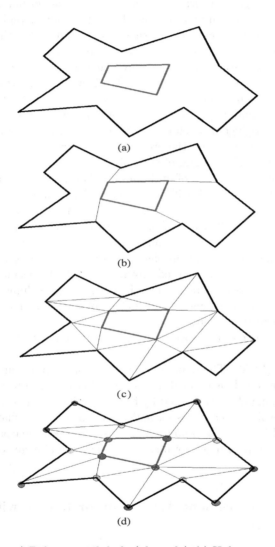

**FIGURE 7.4**: a) Polygon with hole (obstacle), b) Hole connected to polygon through helpers, c) Triangulation of polygon with hole, d) Coloring of polygon with hole. (See ebook for color.)

## 7.7    Camera Placement

In this age of rapid technology growth, camera placement has become a necessity for a really good security system. However, the optimum use of cameras and associated hardware will reduce the cost of any surveillance system. The optimal placement of cameras in surveillance systems originates from a real-world problem of guarding any place where high security is necessary. At airports, railway stations, banks, or any place with the minimum number of guards observing the whole region together is a very important aspect of security. In the computational geometry version of the problem, the layout of the area is represented by a simple polygon and each guard is represented by a point in the polygon, as shown in Figure 7.4. Many researchers have introduced some variations into the problem, such as restricting guards to the perimeter, or even to the vertices of the polygon.

In this section, the problem of optimum number of camera placement, for a given region and vision task will be discussed. Even in a very complicated case, the region itself may change with time, i.e., furniture or walls may be added or removed. The efficiency of any algorithm also depends on the type of camera we are choosing, and designing their layout in the region of interest. The goal in a camera placement problem is to determine the optimal position of cameras for a region to be covered, given a set of task-specific constraints. In the most general case, the region may be an arbitrary volumetric shape, it may be outdoors or indoors, and it may include obstacles. It may even involve dynamic changes in the layout.

Definitions of some of the optical camera parameter definitions are as follows:

1) Field of View (FOV): The maximum angle visible from a camera.

2) Spatial Resolution: It is defined as the ratio between the total number of pixels on its imaging element excited by the projection of a real world object. Higher spatial resolution captures more details and produces sharper images.

3) Depth of Field (DOF): The amount of distance between the nearest and farthest objects that appear in acceptably sharp focus in an image is called DOF.

### 7.7.1    Camera Placement Algorithm for Region with Obstacles

In this section, the camera placement problem with obstacles is discussed. In real life in the given layout, many objects will be placed and may block some part of the area of the camera's FOV. In Figure 7.4 (a), the hole represents objects that will block the FOV of the camera. The hole may be the pillar or any object placed in the given region. First the hole is connected to the polygon through helpers, as shown in Figure 7.4 (b). Since the FOV of any camera is a triangular region, the given space will be partitioned into triangles as shown in Figure 7.4 (c). All the vertices are colored using the *3-coloring technique* such that no vertices of the triangle share the same color, as shown in Figure 7.4 (d) (see ebook for colors). Then the total number of cameras required to be kept on red color vertices are calculated. Similarly, the total number of cameras required to be kept at the yellow and green vertices are calculated. Based on this information, the least number of cameras required is calculated. Nowadays, there are a variety of algorithms available for camera placement. The efficient algorithm will suggest the best placement of cameras, and this placement will minimize your set-up cost and also the maintenance cost of the surveillance system.

## 7.8  Camera Communication

Camera communication refers to the communication process among various camera devices (IP or USB) or systems (a network of cameras) that are linked or networked together to exchange information and share resources. The information sharing allows object parameters to be ferried across multiple cameras. Nowadays, surveillance cameras are increasingly used for capturing images and for communicating information.

A network camera can be configured to send video over an IP network for live viewing and/or recording either continuously, at scheduled times, of an event, or on request from authorized users. Captured images can be streamed as motion JPEG, MPEG-4, or H.264 video using various networking protocols, or uploaded as individual JPEG images using FTP, e-mail, or HTTP (hypertext transfer protocol).

In addition to capturing video, network cameras provide event management and intelligent video functionalities such as video motion detection, audio detection, active tampering alarms, and autotracking. Most network cameras also offer input/output (I/O) ports that enable connections to external devices such as sensors and relays. Other features may include audio capabilities and built-in support for power over ethernet (PoE). Axis network cameras also support advanced security and network management features. Camera communication is also facilitated by picture transfer protocol (PTP). This is a protocol developed by the International Imaging Industry Association to allow the transfer of images from digital cameras to computers and other peripheral devices without the need for additional device drivers.

## 7.9    Multiple Camera Coordination and Cooperation

A single camera for tracking is not useful because of the following reasons:
i) They work in compact areas.
ii) Problems like object segmentation, object occlusion, and so on.
iii) Until recently, multi-camera coordination gave overlapping fields of view.
iv) Multi-camera setup deals with occlusion more effectively.
v) Occlusion can be dealt with as the task of tracking any one person is switched to another camera that can view him clearly.

Multi-camera coordination and control performs [50] the following tasks:
(i) Captures and analyzes videos acquired by multiple cameras.
(ii) Fuses the acquired knowledge from various cameras in the network.
(iii) Performs control actions required in the given surveillance task.

By multi-camera cooperation sensing, the sensors in use not only react based on the sensed data, but also help each other by exchanging information among them. Coupling can be done between two cameras. The main types of coupling are geometric coupling and kinematic coupling. Cooperation between static and dynamic sensors is necessary to analyze low resolution events and even switch to higher resolution events.

In multi-camera systems, there are situations that demand alerting the cameras in the network continuously to properly perform the given surveillance task. For example, if an intruder is suspected, then the cameras should track the intruder continuously. Each camera should be able to acquire video

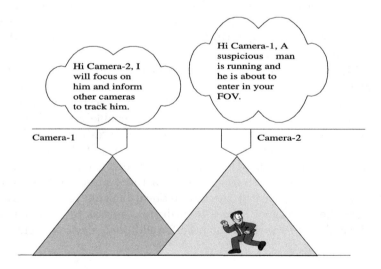

**FIGURE 7.5**: Typical camera coordination scenario.

of the intruder and alert the next camera about the incoming intruder. PTZ cameras can also be instructed to acquire high-resolution views of an intruder, as shown in Figure 7.5, since they have pan tilt and zoom facility.

Also, the cameras currently focusing on the intruder should transmit their data to the concerned neighboring cameras. The cameras receiving the data should be prepared and ready to track the potential intruder. All these operations can only be carried out if there is proper coordination and control among the camera nodes. Analogous to a human security team, where each security person communicates with other security personnel during a policing operation, surveillance cameras are also required to communicate and coordinate with each other.

Along the same line, there should be coordination among all the surveillance cameras. A camera currently focusing on the intruder should transmit its data to the concerned neighboring cameras. The cameras receiving the data should be ready to track the incoming intruder. All these operations can only be carried out if there is proper coordination and control in the camera network.

## 7.10 Summary

In this chapter, the various types of CCTV cameras and their roles for video surveillance have been discussed. The various types of smart cameras are also discussed. The performance of CCD and CMOS imagers is compared and it is observed that CCD imagers performs well for indoor environments while CMOS imagers are suitable for outdoor environments. The demand for smart cameras is continuously increasing, and traditional industries are replacing CCTV cameras with smart cameras, since smart cameras provide intelligent automated surveillance. Application domains such as healthcare, entertainment, and education are exploring the use of smart cameras. Imagers are an emergent solution for practically every automation-focused machine vision application. Camera networking, camera calibration, camera placement, camera communication, and multi-camera coordination are also discussed, to give insight for designing efficient video surveillance systems.

# Chapter 8

# Surveillance Systems and Applications

This chapter begins with a short introduction to surveillance systems, in Section 8.1. A brief overview of video content analytics is presented in Section 8.2. Some application case studies, such as baggage exchange detection and fence crossing detections are discussed in Sections 8.3 and 8.4, respectively. Military applications are discussed in Section 8.5. More details on the use of surveillance systems in transportation are presented in Section 8.6.

## 8.1 Introduction

Video surveillance is of increasing importance as most organizations seek to safeguard physical and capital assets. There is the necessity to observe more people and their activites, and to extract useful information from video data automatically, which demands more scalability, capability, and capacity in video surveillance systems. Installing video cameras is easy, but detecting human activities automatically is still a challenging task. Although surveillance cameras are already in use at banks, shopping malls, airports, railway stations, and parking lots, video data is mostly used for investigation purposes as a forensic tool, thus losing its important benefit as a real-time system. What is expected is continuous monitoring of surveillance video to alert security officers about a suspicious person before he commits a crime.

Video surveillance technology has been used to control traffic, detect accidents on highways, and monitor crowd behavior in public spaces. The numerous military applications include patrolling national borders, monitoring the flow of refugees, and monitoring adherence to peace treaties. The primary goal is maintaining situation awareness by fusing information from multiple sensors into a coherent model of actors, actions, and events to help a remote user understand abnormal and unusal activities.

### 8.1.1 Components of Video Surveillance Systems

A video surveillance system (VSS) basically comprises the following components:

1. Camera and lighting
2. Control and recording unit
3. Output interface and display

The above components are connected using cables, a junction box, or switches. The quality of the above components and their proper connections have an influence on the image quality. Hence, compatibility and quality of all the components should be ensured for successful interaction of the components of a video surveillance system.

## 8.2   Video Content Analytics

Video content analytics (VCA) is the capability to automatically analyze video to detect abnormal activities. This technical capability is used in video surveillance at public places such as airports, railway stations, and banks. The algorithms can be implemented as software on general purpose computers, or as developed dedicated hardware. Many different functionalities can be implemented in VCA, such as video motion detection. More advanced functionalities such as video object tracking and egomotion estimation can be included. It is also possible to build other functionalities in VCA, such as person identification, face detection, behavior analysis, and abnormal activity detection.

### 8.2.1   Functionalities

Several articles provide an overview of the modules involved in the development of video analytic applications. Table 8.1 presents the list of VCA functionalities and their short descriptions.

### 8.2.2   Commercial Applications

VCA is a new technology that has wide applications. Functionalities such as motion detection and people counting are available on the market as commercial products, but still, a lot of research work is required to make the system robust. In many applications, VCA is implemented on CCTV systems, either distributed on the cameras or centralized on dedicated processing systems. Video analytics and smart CCTV are commercial terms for VCA in the security domain. In shopping malls, VCA is used to monitor and track the shoppers from a security point of view. Police and forensic scientists analyze CCTV video when investigating criminal activity. The police use VCA soft-

**TABLE 8.1**: Functions of VCA

| Function | Description |
|---|---|
| Egomotion estimation | It is used to determine the location of a camera by analyzing video sequence. |
| Motion detection | It is used to determine the presence of motion in the observed scene. |
| Shape recognition | It is used to recognize shapes in the input video such as circles, triangles, rectangles. This functionality is typically used for object detection. |
| Object detection | It is used to determine the presence of a type of object, such as person, animal, vehicle. Other examples include fire and smoke detection. |
| Recognition | This functionality is very useful for decision making in surveillance video. Face recognition is used to recognize the person in the scene. Automatic number plate recognition is used to recognize vehicles in the traffic scene. |
| Video object tracking | It is used to determine the location of objects in the given video sequence. |

ware for searching key events and find suspects by performing video content analysis of CCTV footage.

### 8.2.3   Video Object Tracking

This is the process of locating a moving objects over time using a camera. In other words, tracking means associating target objects in consecutive video frames. The tracking of objects can be especially difficult when the objects are moving fast relative to the frame rate. Another situation that increases the complexity is when the tracked object changes orientation over time. In these cases, motion models are usually used to describe how the image of the target might change for different possible motions of the object. It has a variety of uses in application domains such as video object tracking, abnormal activity recgnition, video surveillance, vehicle traffic, human behavior analysis, and crowd behaviour analysis.

## 8.3    Baggage Exchange Detection

In this section, the application of a surveillance system for detecting baggage exchange is presented in detail. In places such as airports, railway stations, and bus stands, the exchange of baggages are observed. As shown in Figure 8.1, the proposed baggage detection algorithm has the following steps:

a. Object detection using GMM

b. Tracking using Kalman filter

c. Labeling of different objects

d. Identification of baggage and person

f. Warning system in case of exchange

### 8.3.1    Object Detection Using GMM

In order to extract moving objects from a video sequence, the current frame is subtracted from the background model. For modeling the background, the GMM-based background model can be used. The background modeling involves developing an algorithm that is able to detect the required object ro-

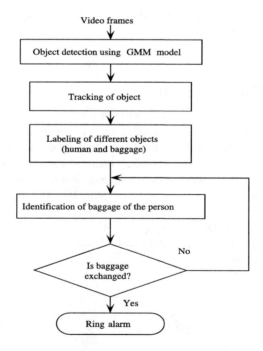

**FIGURE 8.1**: Flow chart for baggage exchange detection

bustly. It should also be able to handle various changes, such as change in the illumination.

### 8.3.1.1 Background Subtraction

The simplest approach to detecting moving objects is to compare background image $B_t(x, y)$ with the current frame $I_t(x, y)$ obtained at the time $t$. The objects can be extracted simply by subtracting each pixel in $B(x, y)$ from the same location as the pixel in $I_t(x, y)$, as shown by Equation 8.1.

$$F_t(x, y) = I_t(x, y) - B_t(x, y) \qquad (8.1)$$

This difference in the image would only show some intensity for the pixel locations that have changed in the two frames. Although we have seemingly removed the background, this approach will only work for cases where all foreground pixels are moving and all background pixels are static. A threshold is put on this image difference to improve the subtraction as $|F_t(x, y) - F_{t-1}(x, y)| > Threshold$. The accuracy of this approach depends on the speed of movement in the scene. Faster movements may require higher thresholds.

**FIGURE 8.2**: This is the case of a person who left his bag, and he, himself, picks up the bag he left. Since this is the normal case, the bounding box color is green. (See ebook for color.)

### 8.3.1.2    Gaussian Mixture Model

In this model, input pixel is compared to the mean $\mu$ of the associated components. If the value of a pixel is close enough to a chosen component's mean, then that component is considered to be the matched component. In order to be a matched component, the difference between the pixels and mean must be less than compared to the component's standard deviation scaled by factor D in the algorithm. Gaussian weight, mean, and standard deviation (variance) are updated to reflect the new obtained pixel value. In relation to non-matched components, the weights $\pi$ decrease whereas the mean and standard deviation stay the same. It is dependent upon the learning component 'p' in relation to how fast they change. Thus it is identified which components are parts of the background model. To do this, a threshold value is applied to the component weights $\pi$. Finally, the foreground pixels are determined. Here the pixels that are identified as foreground do not match with any components determined to be part of the background. Mathematically, GMM is the weighted sum of a number of Gaussians where the weights are determined by a distribution $\pi$.

$$p(x) = \pi_0 N(x|\mu_0, \Sigma_0) + \pi_1 N(x|\mu_1, \Sigma_1) + \ldots + \pi_k N(x|\mu_k, \Sigma_k) \qquad (8.2)$$

**FIGURE 8.3**: This is the case of the unattended bag picked up by another person. As soon as the bag is picked up by another person, the bounding box color changes to red as a warning signal. (See ebook for color.)

where

$$\sum_{i=0}^{k} \pi_i = 1 \tag{8.3}$$

$$p(x) = \sum_{i=0}^{k} \pi_i N(x|\mu_i, \Sigma_i) \tag{8.4}$$

where the $i^{th}$ vector component is characterized by normal distributions with weights $\pi_i$, means $\mu_i$, and covariance matrices $\Sigma_i$.

## 8.3.2 Tracking Using a Kalman Filter

A Kalman filter is used to estimate the state of a linear system where the state is assumed to be distributed as a Gaussian function. The Kalman filter is a recursive predictive filter that is based on the use of state space techniques and recursive algorithms. To improve the estimated state, the Kalman filter uses measurements that are related to the state. Kalman filtering is composed of two steps: prediction and correction.

In the first step, the state is predicted with the dynamic model. The prediction step uses the state model to predict the new state of the variables. Sometimes when a foreground object goes behind an object, we cannot see it. A Kalman filter deals with this type of occlusion problem. It identifies the objects and also determines the predicted path using the particle's dynamic features (like displacement, velocity, and acceleration) and their weight.

## 8.3.3 Labelling of Different Objects

Now we need to label each object. The purpose here is to identify objects with their labels. For labeling, first we find the blobs of different objects. Then we apply bounding blobs around them. For each blob, one label is assigned and the same label is assigned to his/her bag. This helps us to identify the person corresponding to each bag left behind.

## 8.3.4 Identification of Baggage and Person

For labeling purposes, it is necessary to search for bags separately because the tracking model does not differentiate between people and bags. The baggage detection process uses the output of the tracking model and the foreground segmentation for each frame as input, identifies the baggage items, and determines when they are abandoned. The output of the tracking model contains the number of objects, their identities and locations, and parameters of the bounding boxes. The baggage detection process relies on following three important assumptions:

1) Baggage items probably do not move.

2) Baggage items probably appear smaller than people.

3) Baggage items must have an owner.

The algorithm finds the centroid of all the objects present as soon as baggage is detected. At that time, the distance between the centroid of the bag and centroid of the person who left the bag is calculated. If it is less than some threshold value, the identification number (ID) of the person is assigned to that bag.

### 8.3.5   Warning System in Case of Exchange

Now, if the bag is lifted by any other person, then it should show a warning sign. The label corresponding to the person and baggage will help us to find whether the bag belongs to the same person or not. The bag left will have a blob of itself. If a person approaches the bag and if his/her blob gets combined for time $(t > t_0)$, then the system should match his/her IDs. If IDs are matched, then it's all right; otherwise the alarm should get activated.

### 8.3.6   Results

The successful results of two cases are shown in Figure 8.2 and Figure 8.3. Case 1: A person left his bag and he himself picks up the bag. So we do not get any warning here. The snapshot of the output video is shown in Figure 8.2 Case 2: The unattented bag is picked up by another person. We get a warning in this case. The snapshots are shown in Figure 8.3.

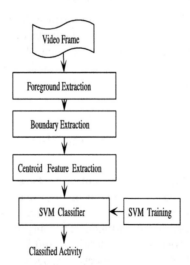

**FIGURE 8.4**: Flow chart of proposed fence-crossing detection approach

## 8.4   Fence-Crossing Detection

Detection of a person climbing a fence is a very important event considering the security of borders, restricted zones, airports, railway stations, shopping malls, and even residential bulidings. Protecting the restricted area surrounded by the fence from an intruder is a very important activity from a surveillance point of view. In [30], Kolekar et al. have presented an approach to automatically detecting the person crossing a fence. A fence-climbing event involves walking, climbing up, and climbing down activities. Figure 8.4 shows the flow chart of the proposed fence-crossing algorithm. First, the moving person is detected by the background subtraction algorithm and a blob is placed around the moving person. Then, the centroid of the blob is computed and its variations along the frames are considered as features. A SVM-based classifier is used for detection of walking, climbing up, and climbing down activities.

### 8.4.1   Proposed Fence-Crossing Detection System

#### 8.4.1.1   Foreground Extraction

Foreground extraction is performed by separating out foreground objects from the background in the video sequence. For any human activity recognition, the first step is the extraction of the moving targets from a given video stream. The segmentation of moving objects of interests from the scene makes our life easier since it reduces computations. Outdoor environments create many problems in activity recognition because of sudden illumination changes, small motion in the background such as tree leaves, and occlusion. The fundamental approach for foreground segmentation is to subtract the given input frame from the estimated background and apply a threshold to the absolute difference to get the foreground region. Estimation of background can be done by frame averaging or GMM [56], [24]. The simplest approach is to use the averaging method of background modeling as follows:

$$B_t(x, y) = \frac{1}{N} \sum_{k=1}^{N} I_{t-k}(x, y) \qquad (8.5)$$

where $B_t(x, y)$ is the background image at time $t$,
$I_t(x, y)$ is the frame at time $t$,
$N$ is the total number of frames.

After estimating the background, it extracts the foreground by subtracting the current background image from the current frame. The foreground extracted image will contain spurious pixels, holes, blurring and many other anomalies. To remove these anomalies from the region of interests (ROI), morphological operations are performed.

### 8.4.1.2    Feature Extraction and Selection

In activity recognition, we select distinct features to differentiate between different activities, such as walking, climbing up, and climbing down. Good feature selection helps in simplifications of the model such that it can be easily interpreted by the users. In the proposed scheme, we have selected five centroid-based features for classifying the activities that are listed in Table 8.2. The centroid of the target is computed as follows:

$$x_c = \frac{1}{N_b} \sum_{i=1}^{N_b} x_i \tag{8.6}$$

$$y_c = \frac{1}{N_b} \sum_{i=1}^{N_b} y_i \tag{8.7}$$

where $x_c, y_c$ is the centroid of target, $N_b$ is the number of boundary points.

**TABLE 8.2**: Features

| Features | Explanation |
|---|---|
| $x_c$ | x-coordinate of the centroid of the detected blob |
| $y_c$ | y-coordinate of the centroid of the detected blob |
| $x_c - y_c$ | Difference between the centroid coordinates |
| $x_{c,1} - x_{c,i}$ | Difference between the x-coordinates of all the centroids of all the target frames. Here i=1,2...T. T is the total number of blobs |
| $y_{c,1} - y_{c,i}$ | Difference between the y-coordinates of all the centroids of all the target frames. Here i=1,2...T. T is the total number of blobs |

### 8.4.1.3    SVM Classifier

An SVM-based classifier is used to classify activities such as walking, climbing up, and climbing down. SVM is a statistical method-based powerful classifier that categorizes the extracted activities into two different classes [31]. In general, it creates a hyperplane or hyperplanes that act as a boundary between all the input classes. The minimum distance between any training data points of two classes is called the margin. Large distance provides a greater margin and the lower generalization error of the classifier. Kernel-based SVM classifiers are used to separate training data, which involves complex geometry of the hyperplane. These classifiers can be defined by

$$f_x = sign \sum_{i=1}^{N_b} x_i \qquad (8.8)$$

The radial basis function kernel (RBF kernel) is used in this case.

## 8.4.2 Experimental Results

Figure 8.5 shows the result of successful detection of a human crossing the fence activity presented in [30]. The proposed method is shape and angle invariant and requires low computation time. It will also be cheaper than a sensor-based detection system. However, this model fails in a multi-person scenario as occlusion causes difficulties in extraction of features. Hence, the proposed algorithm can be made robust by using a Kalman filter for handling the occlusion problem.

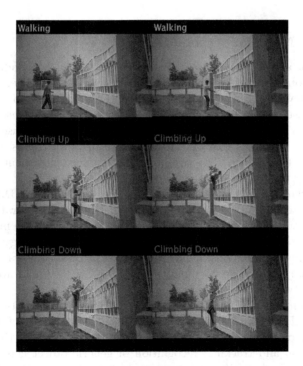

**FIGURE 8.5**: Climbing up and down a fence activity detection result

## 8.5   Military Applications

The applications of video surveillance for military purposes include patrolling national borders, monitoring of the flow of refugees, monitoring adherence to peace treaties, and providing secure perimeters around military bases. During war, these surveillance cameras act as a second pair of eyes on the battlefield, which helps to save the life of many soldiers. Hence the use of CCTV in the military is of extreme importance.

A military base is one of the most monitored places in any country. There are many things to monitor, such as who is coming into military bases and who is going out. Nowadays, aerial cameras are of great importance for military purposes. A camera mounted on the bottom of a drone can be used to acquire images and videos of the battlefield, and helps to save the lives of soldiers. It can also be used as a scouting tool. The drone can fly over a hostile area and inform the troops on the ground of danger. The troops now know what to expect, and the best points of entry and exit to save lives. Hence, CCTV-based surveillance is very useful to the military.

### 8.5.1   The Need for Automation

Nowadays, acquiring a video sequence is very easy, but finding human operators to observe the video screen continuously is expensive. There is a need to automatically collect and extract important activities from battlefield videos and disseminate this information to improve the situational awareness of commanders and staff. Automated video surveillance is very useful in military applications such as providing perimeter security for troops, monitoring whether peace treaties are honored, monitoring refugee movements, and providing security to airports.

In 1997, the Defence Advanced Research Projects Agency (DARPA) Information Systems Office began a program of three years' duration to develop a video surveillance and monitoring (VSAM) system. The main aim of VSAM was to develop automated video understanding technology for battlefield surveillance applications. Under this project, technology was developed to enable a single operator to monitor activities over a broad area using a distributed network of video sensors.

A team composed of the Carnegie Mellon University Robotics Institute and the Sarnoff Corporation were selected to lead the technical efforts for developing a wide range of advanced surveillance techniques: real-time moving object detection and tracking, recognition of generic object classes such as humans, trucks, and specific object types, such as campus police car, object pose estimation with respect to a geospatial site model, active camera control and multi-camera coordination, multi-camera tracking, human gait analysis, and dynamic scene visualization.

## 8.5.2 Basic Design of Surveillance Systems

The basic design suggested in [4] for the use of multisensor surveillance is shown in Figure 8.6. Realistically, it is not possible for a single sensor to monitor a scene considering the cases for limited FOV and the possibility of occlusion of objects that are being monitored. But as every coin has two sides, multisensory surveillance has its own set of challenges, which include how to fuse the information from multiple sensors, how to enable the multiple sensors to track objects cooperatively, and monitoring when further actions should be triggered.

Central to the prototype design is the concept of *smart sensors. Smart sensors* are independently capable of performing real-time detection of objects and their activities. Since the focus of [4] was to eliminate user interaction, there is a requirement of advanced algorithms that fuse sensor data, tasking of sensors to perform autonomous cooperative behaviors, and displaying results in a comprehensible form to the human operator. For this, through a process known as geolocation, every object observation from each sensor is mapped from the camera-centric image-space of the sensor into 3-D geodetic coordinates (latitude, longitude, and elevation). These geolocated objects are then compared with the previous hypotheses, and based on that, the hypotheses are updated.

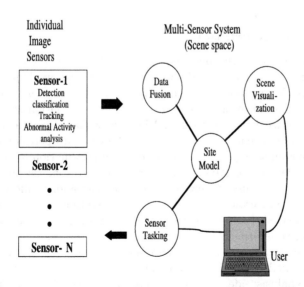

**FIGURE 8.6**: Basic design of surveillance systems for military application

Sensor tasking is done based on an arbitration function that determines the cost of assigning a task to each sensor based on task priority, the load on each sensor, and visibility of the objects from each sensor. The system performs a greedy optimization of the cost to determine which tasking of sensors maximizes the overall system's performance.

## 8.6    Transportation

### 8.6.1    Road Transportation

Nowadays, for controlling traffic, an automatic traffic light controlled by timers and electrical sensors is used. In a traffic light, for each phase is a constant numerical value loaded into the timer. Based on that, the lights are automatically turned on and off. Using electronic sensors is another way to detect vehicles and produce signals. In this method, the time is being wasted by a green light on an empty road. Traffic congestion also occurred while using the electronic sensors for controlling the traffic. To remove these problems, video-based surveillance systems can be used. These systems not only control the traffic, but also detect activities such as collision of vehicles, accidents, and wrong entry by vehicles.

### 8.6.2    Rail Transportation

Maintaining efficient rail operation at the highest levels of customer satisfaction are important benchmarks for modern railways working in a competitive market. Many rail operational areas are already well covered with dedicated telecommunication solutions, passenger information systems, and railway multi-service bearer networks. Video surveillance solutions for railways is a further area that helps railways to achieve their operational targets. Digital video surveillance provides direct access to information relevant to supervising and adapting ongoing railway operation. Automatic processing of this information can be used to provide decision criteria for traffic scheduling, passenger safety, and crowd management inside stations and on platforms. Modern video analysis tools access these video streams and can automatically provide supportive information relevant to disaster prevention, access control for passenger and freight services on board and to ground-based staff. The key benefits of any intelligent video surveillance system for railways are:

i) Intelligent analysis: identification, counting crowd density, automated alarms

ii) Video playback and post-incident analysis

iii) Opening interface architecture to integrate existing surveillance infrastructure

iv) Emergency incident support: Making decisions quickly and handling emergencies efficiently thanks to information obtained from video sources.

### 8.6.3 Maritime Transportation

Over the years, maritime surveillance has become increasingly important because of the increased threat of piracy. While radar systems have been extensively used in maritime environments, these generally require large metallic targets. Modern pirates favor small, fast, non-metallic boats that are difficult to detect. Hence, to detect such boats of pirates, manual detection using dedicated crew members on board is used. Since the attention level of humans is limited, automated video surveillance systems are employed to constantly monitor camera feeds and detect the boats of pirates around the ship. Maritime surveillance has great impact on security, the economy, and the environment. With the growing use of maritime transport, and an increase in pirate attacks, there is a need for automated video surveillance systems to monitor activities such as trafficking of prohibited substances, illegal immigration and fishing, terrorist attacks at port areas, and collisions between marine vehicles. Since about 80% of world trade is carried by sea transport, the attacks against marine vehicles is one way to hurt the economy of a country. Hence, efforts have been made worldwide for the development of maritime surveillance systems.

The European projects AMASS (Autonomous Maritime Surveillance System) [61] and MAAW (Maritime Activity Analysis Workbench) [17] are maritime surveillance systems based on cameras. Many research groups are also working on maritime surveillance systems. Hu et. al [21] present a visual surveillance system for cage aquaculture that automatically detects and tracks intruder ships. Also, a robust foreground detection and background updating technique has been proposed to effectively reduce the influence of sea waves and improve the performance of detection and tracking of ships.

#### 8.6.3.1 Challenges in Maritime Transportation

The main challenges in maritime transportation are as follows:

1) Conventional algorithms for detection and tracking vessels on video do not produce efficient results in a maritime environment, since the background is quite dynamic.

2) The dynamic and unpredictable appearance of the ocean makes detection and tracking a little difficult.

3) The quality of images captured by the cameras may not be good because of environmental conditions such as storms and haze.

4) The white foam on the water surface caused by the waves, the sunlight reflection, and the change in lighting conditions hinder the detector and tracker. 5) The performance of detector and tracker also degrades due to low contrast of the image captured by the cameras and the presence of fog.

---

## 8.7   Summary

In this chapter, various approaches for surveillance systems are discussed. Researchers can use the techniques discussed in this chapter to develop robust systems for various commercial applications. The overall scope of surveillance systems for various domains, such as military applications and transportation, is presented in this chapter.

Video surveillance systems for road and rail transport are discussed in brief in this chapter. Nowadays, maritime video surveillance systems are used to increase coastal and ship security against hostile vessel attacks. They can also be used to control maritime traffic and avoid collisions at ports. The general tracking algorithms seem not to perform well in real situations when little vessels that have low contrast with the background arise in the camera's field of view. The problem of maritime video surveillance has not yet been solved completely, and needs to be explored further.

# Bibliography

[1] Markov A. An example of statistical investigation of the text of *Eugene Onegin* concerning the connections of samples in chains. In *Proceedings of the Academy of Sciences of St. Petersburg*, 1913.

[2] Shobhit Bhatnagar, Deepanway Ghosal, and Maheshkumar H Kolekar. Classification of fashion article images using convolutional neural networks. In *Int Conf on Image Information Processing, India*, pages 1–6, 2017.

[3] Subhomoy Bhattacharyya, Indrajit Chakrabarti, and Maheshkumar H Kolekar. Complexity assisted consistent quality rate control for high resolution h. 264 video conferencing. In *Computer Vision, Pattern Recognition, Image Processing and Graphics (NCVPRIPG), 2015 Fifth National Conference on*, pages 1–4. IEEE, 2015.

[4] Robert T Collins, Alan J Lipton, Takeo Kanade, Hironobu Fujiyoshi, David Duggins, Yanghai Tsin, David Tolliver, Nobuyoshi Enomoto, Osamu Hasegawa, Peter Burt, et al. A system for video surveillance and monitoring. *VSAM final report*, pages 1–68, 2000.

[5] D Comaniciu and P Meer. Mean shift: A robust approach toward feature space analysis. *IEEE Trans. on Pattern Analysis and Machine Intelligence (PAMI)*, 24(5):603–619, 2002.

[6] D Comaniciu, V Ramesh, and P Meer. Real-time tracking of non-rigid objects using mean shift. In *IEEE conf. on Computer Vision and Pattern Recognition (CVPR)*, pages 673–678, 2000.

[7] D Cremers and S Soatto. Motion competition: A variational approach to piecewise parametric motion segmentation. *International Journal of Computer Vision*, 62(3):249–265, 2005.

[8] Ross Cutler and Larry Davis. Robust real-time periodic motion detection, analysis, and applications. *IEEE Transactions on Pattern Analysis and Machine Intelligence*, 22:781–796, 1999.

[9] Deba Prasad Dash and Maheshkumar H Kolekar. A discrete-wavelet-transform-and hidden-markov-model-based approach for epileptic focus localization. *Biomedical Signal and Image Processing in Patient Care*, page 34, 2017.

[10] Jia Deng, Wei Dong, Richard Socher, Li-Jia Li, Kai Li, and Li Fei-Fei. Imagenet: A large-scale hierarchical image database. In *Computer Vision and Pattern Recognition, 2009. CVPR 2009. IEEE Conference on*, pages 248–255. IEEE, 2009.

[11] D Dhane, Maheshkumar H Kolekar, and Priti Patil. Adaptive image enhancement and accelerated key frame selection for echocardiogram images. *Journal of Medical Imaging and Health Informatics*, 2(2):195–199, 2012.

[12] Baum L E, Petrie T, Soules G, and Weiss N. A maximization technique occuring in the statistical analysis of probablistic functions of Markov chains. In *Annals of Mathematical Statistics*, volume 41, pages 164–171, 1970.

[13] Y Freund and R E Schapire. A decision-theoretic generalization of online learning and an application to boosting. *Journal of Computer and System Sciences*, 55(1):119–139, 1997.

[14] Flohr T G, Schaller S, Stierstorfer K, Bruder H, Ohnesorge B M, and Schoepf U J. Multi-detector row ct systems and image-reconstruction techniques. *Radiology*, 135(3):756–773, 2005.

[15] Leon A Gatys, Alexander S Ecker, and Matthias Bethge. A neural algorithm of artistic style. *arXiv preprint arXiv:1508.06576*, 2015.

[16] Ross Girshick, Jeff Donahue, Trevor Darrell, and Jitendra Malik. Rich feature hierarchies for accurate object detection and semantic segmentation. In *The IEEE Conference on Computer Vision and Pattern Recognition (CVPR)*, June 2014.

[17] Kalyan M Gupta, David W Aha, Ralph Hartley, and Philip G Moore. Adaptive maritime video surveillance. Technical report, Knexus Research Corp, Springfield, VA, 2009.

[18] Bing Han, William Roberts, Dapeng Wu, and Jian Li. Robust feature-based object tracking. In *Proc. of SPIE Vol*, volume 6568, pages 65680U–1, 2007.

[19] B K Horn and B G Schunck. Determining optical flow. *Cambridge, MA, USA, Technical Report*, pages 781–796, 1980.

[20] Weiming Hu, T. Tan, L. Wang, and S. Maybank. A survey on visual surveillance of object motion and behaviors. *IEEE Transactions on Systems, Man and Cybernetics, Part C*, 34(3):334–352, 2004.

[21] Wu-Chih Hu, Ching-Yu Yang, and Deng-Yuan Huang. Robust real-time ship detection and tracking for visual surveillance of cage aquaculture. *Journal of Visual Communication and Image Representation*, 22(6):543–556, 2011.

[22] Jaideep Jeyakar, R Venkatesh Babu, and KR Ramakrishnan. Robust object tracking with background-weighted local kernels. *Computer Vision and Image Understanding*, 112(3):296–309, 2008.

[23] Hu M K Visual pattern recognition by moment invariants. In *IRE Transaction on Information Theory*, pages 179–187. IT 8, 1962.

[24] P Kaewtrakulpong and R Bowden. An improved adaptive background mixture model for real-time tracking with shadow detection. *Video-based surveillance systems*, pages 135–144, 2002.

[25] M Kilger. A shadow handler in a video-based real-time traffic monitoring system. In *IEEE Workshop on Applications of Computer Vision*, pages 11–18, 1992.

[26] Maheshkumar H Kolekar. An algorithm for designing optimal Gabor filter for segmenting multi-textured images. *IETE Journal of Research*, 48(3-4):181–187, 2002.

[27] Maheshkumar H Kolekar. Hidden markov model based highlight generation of cricket video sequence using video and audio cues. *CSI Communications*, 28(7):25–26, 2005.

[28] Maheshkumar H Kolekar. Bayesian belief network based broadcast sports video indexing. *Int. Journal on Multimedia Tools and Applications*, 54(1):27–54, 2011.

[29] Maheshkumar H Kolekar and U Rajendra Acharya. A special section on biomedical signal and image processing. *Journal of Medical Imaging and Health Informatics*, 2(2):147–148, 2012.

[30] Maheshkumar H Kolekar, Nishant Bharati, and Priti N Patil. Detection of fence climbing using activity recognition by support vector machine classifier. In *IEEE Region 10 Int. Conf.(TENCON)*, pages 1–6, 2016.

[31] Maheshkumar H Kolekar, D P Dash, and Priti N Patil. Support vector machine based extraction of crime information in human brain using erp image. In *Int Conf on Computer Vision and Image Processing*, pages 1–6, 2016.

[32] Maheshkumar H Kolekar and Deba Prasad Dash. Hidden Markov model based human activity recognition using shape and optical flow based features. In *Region 10 Conference (TENCON), 2016 IEEE*, pages 393–397. IEEE, 2016.

[33] Maheshkumar H Kolekar and Vinod Kumar. *Biomedical Signal and Image Processing in Patient Care*. IGI Global, 2017.

[34] Maheshkumar H Kolekar, K Palaniappan, S Sengupta, and G Seethara-man. Semantic concept mining based on hierarchical event detection for soccer video indexing. *Int. Journal on Multimedia*, 4(5):298–312, 2009.

[35] Maheshkumar H Kolekar and S Sengupta. Hidden markov model based structuring of cricket video sequences using motion and color features. In *Indian Conf. on Computer Vision, Graphics and Image Processing*, pages 632–637, 2004.

[36] Maheshkumar H Kolekar and S Sengupta. Hidden markov model based video indexing with discrete cosine transform as a likelihood function. In *India Annual Conference, 2004. Proceedings of the IEEE INDICON 2004. First*, pages 157–159. IEEE, 2004.

[37] Maheshkumar H Kolekar and S Sengupta. Semantic indexing of news video sequences: a multimodal hierarchical approach based on hidden markov model. In *IEEE Region 10 Int. Conf.(TENCON)*, pages 1–6, 2005.

[38] Maheshkumar H Kolekar and S Sengupta. Event importance based cus-tomized and automatic cricket highlight generation. In *IEEE Int. Conf. on Multimedia and Expo (ICME)*, volume 61, pages 1617–1620, 2006.

[39] Maheshkumar H Kolekar and S Sengupta. Semantic concept mining in cricket videos for automated highlight generation. *Int. Journal on Mul-timedia Tools and Applications*, 47(3):545–579, 2010.

[40] Maheshkumar H Kolekar and S Sengupta. Bayesian network-based cus-tomized highlight generation for broadcast soccer videos. *IEEE Trans-action on Broadcasting*, 61(2):195–209, 2015.

[41] Maheshkumar H Kolekar, S N Talbar, and T R Sontakke. Texture seg-mentation using fractal signature. *IETE Journal of Research*, 46(5):319–323, 2000.

[42] Alex Krizhevsky, Ilya Sutskever, and Geoffrey E Hinton. Imagenet classi-fication with deep convolutional neural networks. In *Advances in neural information processing systems*, pages 1097–1105, 2012.

[43] M P Kumar, P H Torr, and A Zisserman. Learning layered motion segmentations of video. *International Journal of Computer Vision*, 76(3):301–319, 2008.

[44] C T Lin, C T Yang, Y W Shou, and T K Shen. An efficient and robust moving shadow removal algorithm and its applications in its. *EURASIP Journal on Advances in Signal Processing*, pages 1–19, 2010.

[45] Tsung-Yi Lin, Michael Maire, Serge Belongie, James Hays, Pietro Perona, Deva Ramanan, Piotr Dollár, and C Lawrence Zitnick. Microsoft COCO:

Common objects in context. In *European conference on computer vision*, pages 740–755. Springer, 2014.

[46] Tang S Ling, Liang K Meng, Lim M Kuan, Zulaikha Kadim, and Ahmed A. Colour-based object tracking in surveillance application. In *Int Multiconference of Engineers and Computer Scientists*, pages 11–18, 2009.

[47] A J Lipton, H Fujiyoshi, and R S Patil. Moving target classification and tracking from real-time video. In *IEEE Workshop on Applications of Computer Vision*, pages 8–14, 1998.

[48] B D Lucas and T Kanade. An iterative image registration technique with an application to stereo vision. In *International Joint Conference on Artificial Intelligence*, pages 674–679, 1981.

[49] B D Lucas and T Kanade. An iterative image registration technique with an application to stereo vision. In *Proceedings of Image Understanding Workshop*, pages 121–130, 1981.

[50] Prabhu Natarajan, Pradeep K Atrey, and Mohan Kankanhalli. Multi-camera coordination and control in surveillance systems: A survey. *ACM Transactions on Multimedia Computing, Communications, and Applications (TOMM)*, 11(4):57, 2015.

[51] H Rai, Maheshkumar H Kolekar, N Keshav, and J K Mukherjee. Trajectory based unusual human movement identification for video surveillance system. In *Progress in Systems Engineering*, pages 789–794. Springer, 2015.

[52] C Rasmussen and G D Hager. Probabilistic data association methods for tracking complex visual objects. *IEEE Transactions on Pattern Analysis and Machine Intelligence*, 23(6):560–576, 2001.

[53] Samsu Sempena, Nur Ulfa Maulidevi, and Peb Ruswono Aryan. Human action recognition using dynamic time warping. In *Electrical Engineering and Informatics (ICEEI), 2011 International Conference on*, pages 1–5. IEEE, 2011.

[54] J Shi and C Tomasi. Good features to track. In *IEEE Int. Conf. on Computer Vision and Pattern Recognition*, pages 593–600, 1994.

[55] E P Simoncelli and W T Freeman. The steerable pyramid: A flexible architecture for multi-scale derivative computation. In *IEEE International Conference on Image Processing*, pages 444 – 447, 1995.

[56] Chris Stauffer and W E L Grimson. Adaptive background mixture models for real-time tracking. In *IEEE Int Conf Computer Vision and Pattern Recgnition*, volume 2, pages 246–252, 1999.

[57] Christian Szegedy, Sergey Ioffe, Vincent Vanhoucke, and Alexander A Alemi. Inception-v4, inception-resnet and the impact of residual connections on learning. In *AAAI*, pages 4278–4284, 2017.

[58] C Tomasi and T Kanade. Factoring image sequences into shape and motion. In *Proceedings of the IEEE Workshop on Visual Motion*, pages 21–28, 1991.

[59] N Vaswani, A Tannenbaum, and A Yezzi. Tracking deforming objects using particle filtering for geometric active contours. *IEEE Transactions on Pattern Analysis and Machine Intelligence*, 29(8):1470–1475, 2007.

[60] P Viola and M J Jones. Robust real-time face detection. *International Journal of Computer Vision*, 57(2):137–154, 2004.

[61] K Wolfgang and O Zigmund. Robust layer-based boat detection and multi-target-tracking in maritime environments. In *Waterside Security Conference (WSS), 2010 International*, pages 1–7. IEEE, 2010.

[62] Christopher Wren, Ali Azarbayejani, Trevor Darrell, and Alex Pentland. Pfinder: Real-time tracking of the human body. In *IEEE Transactions on Pattern Analysis and Machine Intelligence*, volume 19, pages 780–785. IEEE, 1997.

[63] Jean Yves Bouguet. Pyramidal implementation of the affine Lucas Kanade feature tracker description of the algorithm. In *Intel Corporation*, volume 4, pages 1–10, 2001.

[64] J Zhang, F Shi, J Wang, and Y Liu. 3d motion segmentation from straightline optical flow. *Multimedia Content Analysis and Mining*, pages 85–94, 2007.

[65] W Zhang, X Z Fang, X K Yang, and Q M J Wu. Moving cast shadows detection using ratio edge. *IEEE Transactions on Multimedia*, 9(6):1202–1214, 2007.

[66] Ye Zhang and Zhi-Jing Liu. Irregular behavior recognition based on treading track. In *IEEE Int. Conf. on Wavelet Analysis and Pattern Recognition*, pages 1322–1326, 2007.

# Index

PGMO 07/10/2018